U.S. Marine Corps Physical Readiness Training for Combat

Plus

Marine Pistol Marksmanship

Department of Defense

Pentagon
Publishing

UNITED STATES MARINE CORPS
Marine Corps Combat Development Command
Quantico, Virginia 22134-5001

29 January 1988

FLEET MARINE FORCE REFERENCE PUBLICATION............0-1B

MARINE PHYSICAL READINESS TRAINING FOR COMBAT

1. PURPOSE

Fleet Marine Force Reference Publication (FMFRP) 0-1Marine Physical Readiness Training for Combat is intended for use by all Marines. It provides the information and references necessary to establish and conduct physical conditioning programs to prepare Marines for the physical demands of combat.

2. APPLICABILITY

Although not directive, this handbook is intended for use as a reference by all Marines in developing unit physical readiness training for combat. This FMFRP does not establish policy for the conduct of physical fitness training. Refer to Marine Corps Order 6300.3_ for applicable Marine Corps policy on physical fitness. It may also be used as a basis for instruction conducted by unit leaders or as a reference for instruction or training.

3. RECOMMENDATIONS

To improve this manual, the MAGTF Warfighting Center seeks comments and recommendations from the user. Officers and noncommissioned officers with constructive comments are encouraged to respond. Lack of response will be considered as concurrence. The User Suggestion Form which follows can be detached and forwarded to - -

Commanding General
Marine Corps Combat Development Command (WF 06)
Quantico, Virginia 22134-5001

4. DISTRIBUTION

Initial distribution of this FMFRP 0-1B is shown below. Addressees on this list should ensure that appropriate subordinate organizations are provided copies.

JOHN I. HOPKINS
Brigadier General, U.S. Marine Corps
Assistant Chief of Staff, Warfighting
Marine Corps Combat Development Command
Quantico, Virginia

DISTRIBUTION: TDJ

From:

To: Commanding General, Marine Corps Combat Development Command (WF 06), Quantico, Virginia 22134-5001

Subj: FMFRP 0-1B, MARINE PHYSICAL READINESS TRAINING FOR COMBAT, RECOMMENDATIONS CONCERNING

1. In accordance with the Promulgation Page to FMFRP 0-1B, which invites individuals to submit suggestions concerning this FMFRP directly to the above addressee, the following unclassified recommendation is forwarded:

Page________Article/Para No._________Line No.________Figure/Table No._________

Nature of Change:_____Add_____Delete_____Change_____Correct

2. Proposed New Verbatim Text: (Verbatim, double spaced; continue on additional pages as necessary.)

3. Justification/Source: (Need not be double spaced.)

NOTE: Only one recommendation per page.

PREFACE

MARINE PHYSICAL READINESS TRAINING FOR COMBAT

Chapter 1
PHYSICAL READINESS LEADERSHIP
Section I. Role of Physical Fitness in Combat

1101. PURPOSE OF PHYSICAL FITNESS TRAINING

Physical fitness training in the Marine Corps has one purpose: to prepare Marines to physically withstand the rigors of combat. All other goals of physical fitness training are subordinate to and must support attainment of this goal. The idea that only infantry or reconnaissance units and their attachments normally face physically demanding combat is wrong. This error must not influence the priority commanders of combat support, combat service support, aviation, and headquarters units give to physical fitness for combat. Physical fitness for combat has a high priority for all Marines. A sound, effective unit program, requiring limited time and material, offers a greater payoff in combat than many more expensive and time-consuming training programs. This handbook describes the concept and provides the ingredients of a program for commanders to use in carrying out one of their most serious responsibilities: ensuring that their Marines are physically ready for combat.

1102. SCOPE OF THE MANUAL

This handbook provides guidance for all leaders, trainers, and planners of physical training programs. It describes unit physical fitness for combat training in the following chapters:

a. **Physical Readiness Leadership.** Chapter 1 provides guidance to leaders in the conduct of physical combat readiness training.

b. **Physical Readiness Training Programs.** Chapter 2 discusses how to structure programs to meet physical readiness goals in a variety of combat training situations.

c. **Physical Conditioning Activities.** Chapter 3 discusses the primary physical conditioning activities which commanders may use to attain readiness for combat goals. Sections in this chapter address foot marches under load, strength-building activities, and activities which build confidence and the aggressive spirit needed in combat.

d. **Combat Water Survival.** As an amphibious force, Marines cannot ignore the importance of combat water survival training. Chapter 4 describes a program for commanders to conduct this training.

e. **Competitive Conditioning Activities.** Chapter 5 describes competitive activities which stress the relationship to combat actions, featuring team-building types of competition.

f. **Evaluation of Performance During Training.** Chapter 6 describes tests which measure individual and unit physical fitness for combat.

g. **The Human Body.** Chapter 7 covers the structure and functioning of the body.

1103. COMMANDER'S ROLE

Major General Commandant John A. Lejeune in the 1921 edition of Marine Corps Manual expressed our philosophy of leadership. Among other things, General Lejeune stated that "it will necessary for officers ... to endeavor to enlist the interest of their men in building up and maintaining their bodies in the finest physical condition." It is significant that the Commandant prefaced these writings in 1921 with the statement that his thoughts were greatly influenced by the experience of World War I. We have never varied as a Corps from this belief that one of a commander's most serious responsibilities is to have Marines physically prepared for combat. Senior commanders and their staffs carry out this role primarily by providing command emphasis, including planning, support, and supervision. Commanders at the battalion/squadron and company/battery level execute a program of physical fitness for combat by close supervision and, most importantly, leadership by example.

1104. HISTORY OF MILITARY PHYSICAL READINESS

Every war has revealed our military physical deficiencies during the initial periods of mobilization. This realization followed the Civil War and has recurred regularly with each national emergency.

a. Training programs in each war were geared to the physical need of the era. Success was dependent upon the amount of time available during training to physically prepare Marines for battle conditions. Frequently, casualties in initial engagements were attributed to the inability of our Marines to physically withstand the rigors of combat over rugged terrain and under unfavorable climatic conditions. With adequate preparation, Marines have always handled the test of battle.

b. During World War II, the first physical conditioning doctrine that could be scientifically justified by testing was introduced. As the war progressed, this program was effective in the physical conditioning of millions of Marines for combat.

c. Postwar periods have traditionally been a time of consolidation. Unfortunately, some leaders considered the conditioning phase of training to be a wartime tool. With such a philosophy prevalent between wars, physical readiness was relegated to secondary importance resulting in a serious lowering of combat effectiveness. The initial commitment of Marines in Korea dramatically displayed this failure to recognize the extreme physical demands of warfare. Call-up of large numbers of reserve Marines and pressing them into the war within weeks after call-up mainly caused the loss in effectiveness.

d. Costly lessons learned from our military experiences over a period of years and the course of several wars led to an increasing interest in the physical conditioning of the individual Marine. No longer can we afford to emphasize physical fitness during wartime and de-emphasize it during peacetime. It is evident that, in spite of increased mechanization and modern weapons, physical readiness retains a vital place in the life of each individual Marine and in every unit within the Marine Corps regardless of mission.

e. Commanders are well aware of the need for rugged and well conditioned Marines. The daily demands of housekeeping, maintenance, support, training, operations, and other time-consuming tasks make it necessary for commanders to set aside time for frequent, regular, and vigorous exercise periods.

1105. PHYSICAL DEMANDS OF COMBAT

There are three primary elements of effective physical fitness for combat: lower body strength and stamina; upper body strength and stamina; and a competitive, combative spirit. Cardiovascular functioning is not treated as a separate element since it is a necessary foundation for the other three. Exercises and activities designed to improve Marines' condition in the three primary elements will foster good cardiovascular and respiratory condition.

a. Lower Body Strength and Stamina. Some common demands that combat places on lower body strength and endurance are marching long distances under load and functioning effectively at the destination; moving quickly and evasively under fire; and carrying wounded Marines to safety.

b. Upper Body Strength and Stamina. Some common demands that combat places on upper body strength and stamina are rapidly emplacing crew-served weapons; handling large-caliber ammunition for extended periods; climbing walls, cliffs, and other high obstacles; and performing field maintenance on air craft or heavy machinery.

c. Competitive, Combative Spirit. While part of this critical aspect of combat readiness is mental and emotional, robust physical condition and a training program which promotes physical aggressiveness greatly enhance a competitive and combative spirit. Activities which force Marines to overcome natural physical fear, which directly leads to fatigue, are particularly valuable.

1106. PHYSICAL EFFECTS OF COMBAT STRESS

It has been thoroughly documented that the added uncertainty and stress of combat have a major physical effect on Marines as well as the acknowledged psychological effect. In other words, in combat, fear equals fatigue. In training, we cannot easily reproduce this combat stress which reduces the effectiveness of individuals and units. However, we can produce fatigue and psychological doubt in training situations by developing a challenging physical training program which stresses the participants. This compels Marines to exercise their ability to continue to function under difficult and trying conditions. Such activities as long foot marches under load and difficult confidence courses train Marines to overcome their own fear and fatigue. Rugged competitive activities such as martial arts and pugil stick fighting prepare Marines psychologically to overcome an opponent as well as their own fear and fatigue. This sort of physically demanding training builds self-confidence and unit morale. The adage that "the unit that sweats more in peacetime will bleed less in battle" certainly applies to its physical training program. Athletic coaches have long followed this approach by such practices as having teams shoot foul shots after wind sprints in order to strengthen their mental conditioning and toughness. Although pushing Marines to their limit is necessary, commanders often set lower standards initially than ultimately desired. The program should then call for steadily increasing the difficulty. A good guide for a starting point is a standard that will be physically and psychologically demanding for no less than three-fourths of the unit. Expressed another way, no more than one-fourth of the unit should be working at less than their limit. The program should progress quickly, augmented, if needed, by remedial training for those unable to meet the unit standards, until the training is challenging for all but the most exceptional Marines in the command. Commanders must never forget that Marines expect to be challenged and thrive on being pushed to their limit.

1107. LIMITS OF PHYSICAL READINESS

A well-conditioned Marine unit has a significant advantage in combat, but this does not mean that physical conditioning can substitute for technical knowledge or good planning. For instance, units must, as discussed in chapter 3, pursue a continuing and challenging program of marches under loads. This does not disprove the axiom that fighting men, if required to carry over one-third of body weight, will generally become ineffective in battle. Leaders and planners must decide what is essential to be carried and must use an effective logistic distribution system, such as unit trains, to make remaining items available when they are needed. In a similar vein, the valid requirement to foster a competitive and combative spirit in Marines does not excuse leaders from the responsibility to practice tactics which will not needlessly risk lives. Physical readiness for combat is a vital goal, but it is only one part of a unit's overall readiness for the lest of combat.

Section II. Fundamentals of Physical Fitness

1201. FITNESS FOR MARINES

Total combat readiness includes technical proficiency and mental, emotional, and physical fitness. If any one of these attributes is lacking, combat effectiveness suffers proportionately. Without technical fitness, Marines lack the knowledge and skill to fight; without mental and emotional fitness, they lack the incentive and will to fight; and without physical fitness, they lack the physical ability and confidence to fight. Physical fitness in a Marine means a whole and healthy body, the capacity for skillful and sustained performance, the ability to recover from exertion rapidly, the desire to complete a designated task, and the confidence to face any eventuality.

1202. COMPONENTS OF PHYSICAL FITNESS

A sound body, free of disease and defect, does not in itself constitute physical fitness. Before an untrained Marine can be considered physically fit for combat, the following traits must be developed:

a. Strength. Every Marine needs enough strength to perform the heaviest task encountered in routine and emergency activities. The basic areas where strength is required are in the arm and shoulder girdle, abdomen, back, and legs. Muscles increase in size, strength, and firmness with regular and strenuous exercises. Without work, they grow flabby and weak.

b. Endurance. Each Marine needs sustaining power to maintain the maximum ability without undue fatigue. There are two types of endurance:

(1) Muscular Endurance. The Marine needs muscular endurance to fight the enemy under the most tiring combat conditions. Muscular endurance is the ability to perform continuous work over long periods of time. Endurance depends on the bloodstream's ability to deliver large amounts of oxygen and nutrition to the muscle masses and then carry away the waste products quickly.

(2) Cardiovascular-Respiratory Endurance. The development of cardiovascular-respiratory endurance ("wind") is necessary to maintain muscular endurance. Cardiovascular-respiratory endurance depends on the efficiency of the blood vessels, lungs, and heart. The maximum effort a Marine can exert over a period of time is limited by the capacity to absorb oxygen and expel carbon dioxide. The average Marine's cardiovascular-respiratory capacity can be greatly increased by exercise.

c. Agility. A Marine must be able to change direction quickly and as faultlessly as possible. The ability to react instantly and to maintain orientation during rapid changes of body position is important to survival. This important ability may be developed by conditioning exercises that require varied and rapid changes of body position on the ground and in the air, such as obstacle courses.

d. Coordination. Coordination is the ability to move all parts of the body in a smooth, efficient, and concerted effort (commonly called timing). A well-coordinated individual does not make useless movements. An individual who moves with precision and accuracy saves energy. Coordination is best developed by practicing diversified muscular activities and skills affecting all body parts.

1203. TYPES OF EXERCISE

Basically, there are two forms of exercise: isotonic and isometric. Both forms are contained in the activities found in this manual.

a. Isotonic. Isotonic exercises are those in which the expenditure of energy is regulated and released during concentrated efforts. The regulated expenditure of energy is controlled by both the mode of exercise and the individual's effort. This type of effort is common to most exercise and sports. To develop endurance, coordination, and flexibility, isotonic exercise must be used. Strength can also be developed or increased through this type of exercise.

b. Isometric. Isometric exercises are those in which maximum effort is applied and held until the engaged muscle energy is depleted during a single contraction effort. The individual exerts full force against an immovable object for a relatively short period of time (6 to 10 seconds) and then repeats the contraction several times with a short period of rest between each contraction. This type of exercise develops only strength; therefore, it has limited application.

1204. PRINCIPLES OF PHYSICAL CONDITIONING

Men and women vary in their physical makeup. Physiological (body) function and reaction also vary in proportion to heavy demands placed upon the body. To attain the maximum benefit without sacrificing Marines' welfare, fitness must grow with a careful program of conditioning. To allow for adjustments in body functioning as the conditioning program progresses and to ensure attainment of objectives, the following principles of physical conditioning must be applied:

a. Overload. As strength and endurance increase, the physical load must be increased until the desired level of fitness is reached.

b. Progression. In beginning stages, the load must be moderate. Gradual progression from this low state of fitness to a higher state is possible through application of a progressive program.

c. Balance. An effective program utilizes various types of activities and provides for the concurrent development of strength, endurance, and coordination as well as basic physical skills.

d. Variety. Some programs fail because the routine becomes boring. The most successful programs always include conditioning activities, competitive events, and military physical skill development.

e. Regularity. There is no easy or occasional way to develop physical fitness. Regularity of exercise is a must, with daily exercise preferred.

1205. THREE STAGES OF PHYSICAL CONDITIONING

Unconditioned or poorly conditioned Marines pass through the following stages in reaching the desired state of physical condition.

a. Toughening Stage. This stage is approximately 2 weeks long and is usually characterized by muscular stiffness and soreness followed by recovery.

b. Slow Improvement Stage. This stage is approximately 6 to 10 weeks long and is characterized by slow and steady improvement until the desired level, or a high level, of fitness is attained.

c. Sustaining Stage. This stage goes on indefinitely in order to maintain the level of conditioning achieved by passage through the previous stages.

1206. EFFECTS OF CLIMATIC CONDITIONS

Temperature, both atmospheric and body, affects the physical performance of Marines. Proper maintenance of body temperature through warm up exercise, proper dress in cold weather, and removal or adjustment of clothing in hot weather is necessary for effective performance and health. Climatic factors to be considered are the following:

a. Exercise In High Temperatures and High Humidity. Marines can endure strenuous physical activity in extremely hot temperatures if they are given an opportunity to become acclimated and if they take enough salt and water. It is essential to continue physical training programs in hot climates. Marines can better withstand high temperatures when they are well-conditioned. High humidity combined with high temperatures presents a serious danger. These conditions prevent the natural cooling of the body by the evaporation of perspiration. Training schedules should conform to the provisions of the base commander in respect to wet bulb conditions. Those who conduct training under these conditions should monitor weight loss and be careful to make fluid replacement adjustments.

b. Exercise at High Altitudes. Certain problems are encountered in conditioning Marines stationed in high altitudes because the heart undergoes greater exertion during exercise. It is particularly important that only light exercise be given initially at such altitudes. A Marine's body gradually adjusts to high altitudes within a few weeks. After this adjustment, progressively greater amounts of exercise are possible.

c. Exercise in Arctic Regions. Military duty in the arctic is so arduous that a high level of physical conditioning is essential. Because of the difficulties of carrying on physical conditioning exercises in extreme cold, Marines should be conditioned to the highest level possible before they arrive. A sustaining program will then maintain that level. When exercising in cold weather, Marines should be required to remove excess clothing to prevent them from becoming damp with perspiration.

1207. WARMING UP AND COOLING OFF

It is a fundamental physiological principle that Marines should warm up gradually before taking strenuous exercise. Such action speeds up the circulation to prepare the body to take an overload and helps to prevent injury to muscles and joints after exercising. Marines should be kept mildly active, walking, stretching, or performing some other mild muscular activity until breathing and temperature have returned to normal. Marines should never be allowed to cool off too rapidly; in cool or cold weather, they should put on additional clothing during the cooling-off period.

1208. PHYSICAL ACTIVITY AS AGE INCREASES

Combat makes severe physical demands on Marines. All Marines, regardless of age, must be physically ready to meet these demands. There is no physiological reason to cease exercise or exertion with age. Increased age usually brings increased responsibility which, in many instances, leads to a routine that can become almost devoid of physical activity. The key to fitness with increased age is to continue exercising at a reasonable level and to include exercise of a vigorous type in the daily routine. Older persons who have not regularly maintained a reasonable state of physical fitness require more time than younger persons to become fit. Such individuals usually require a longer period of time to recover from physical effort than younger Marines. If general health is good, evidence shows that older Marines can develop and maintain a degree of fitness which permits vigorous activity and proper performance of duties. It is both the individual's and the commander's responsibility that all Marines continue a daily sustaining exercise program.

Section III. Goals of Physical Readiness Training

1301. NECESSITY FOR PHYSICAL READINESS TRAINING

A very important objective of training is attainment and maintenance of operational readiness. Marines must be physically ready for operations at any time and under all conditions of climate and environment. A combination of training to develop proficiency in physical skills and conditioning to improve strength and endurance results in physical readiness for combat. The degree of physical fitness required of Marines can be acquired through physical exercises in a normal well-coordinated and closely monitored training routine. The performance of purely military duties, such as drills and marching, is not enough to build all the desired areas of fitness. Few recruits are physically fit for the arduous duties ahead of them. The softening influences of our mechanized civilization and the difficulties of conditioning Marines make physical fitness more important than ever before. If Marines are to be fully developed to and maintained at the desired standard of physical fitness, a well-conceived plan of physical readiness training must be a basic part of every training program. Marines cannot be adequately prepared in any other way for the hard work and arduous demands required on the battlefield.

1302. OBJECTIVE OF PHYSICAL READINESS TRAINING

The overall objective of the physical readiness training program is to develop individuals and units who are physically able and ready to perform their duty assignments or missions during training and in combat. To attain the objective of physical readiness, exercise activities must develop the following areas:

a. Strength and Endurance. Developing adequate strength to perform required duties and adequate endurance to sustain activity over a long period of time.

b. Muscle Tone. Developing muscle tone adequate to maintain proper posture and reasonable weight control.

c. Skills. Developing military physical skills which are essential to personal safety and effective combat performance. As skill is developed, agility and coordination are attained. The essential skills are-

(1) **Marching Under Load.** Marching with individual and unit weapons and equipment.

(2) **Running.** Distance and sprint running on roads and cross-country.

(3) **Jumping.** Broad jumping and vertical jumping downward from a height.

(4) **Dodging.** Changing body direction rapidly while running.

(5) **Climbing and Traversing.** Vertical climbing of ropes, poles, walls, and cargo nets. Traversing horizontal objects such as ropes, pipes, and ladders.

(6) **Crawling.** High crawling and low crawling for speed and stealth.

(7) **Throwing.** Propelling objects, such as grenades, for a distance with accuracy.

(8) **Vaulting.** Surmounting low objects, such as fences and barriers, by use of hand assists.

(9) **Carrying.** Carrying objects and employing man-carries.

(10) **Balancing.** Maintaining proper body balance on narrow walkways and at heights above normal.

(11) **Falling.** Contacting the ground from standing, running, and jumping postures.

(12) **Surviving in Water (Or Other Specialized Situations)** Using water survival techniques.

Character Traits. Instilling character traits which help accomplish military missions to include-

(1) **Confidence.** Confidence develops through achieving progressively more difficult tasks as physical ability grows.

(2) **Aggressiveness.** Participation in combative activities and contests develops desire and willingness to overcome an opponent.

(3) **Reaction Under Pressure.** Competitive contests and game situations train Marines to think and to act quickly while under pressure.

(4) **Teamwork.** Teamwork develops through competitive events in which a number of Marines must coordinate their efforts to accomplish a physical task.

1303. BENEFITS OF EXERCISE

The benefits of exercise are not always understood. Some of the more important results of exercise are listed below:

a. Improved Muscle Tone. Muscular tone improves and, at the same time, muscular strength and endurance are built up.

b. Cardiovascular-respiratory Endurance. Cardiovascular-respiratory endurance, or wind, improves through a process of opening up dormant lung capacity to absorb greater amounts of oxygen.

c. Circulation. Circulation of the blood speeds up and extends to a greater portion of the body as exercise forces the blood to service all parts of the body. The efficiency and effectiveness of the heart, lungs, and blood vessels improve.

d. Flexibility. A wider range of muscular movement is possible and rapidity in physical skills grows.

e. Elimination of Body Waste. Bending and twisting the body and the general speedup of body processes caused by exercise regulate and help eliminate body wastes.

f. Tension. Working off excess nervous energy and relief from daily worries and cares relieve tension. Participation in exercises leaves little time for worry.

g. Sleep. Sleep improves because muscles are healthfully tired after about of exercise. A by-product of sound sleep is relief of tension.

h. Obesity Control.. Control of obesity (fat) is made possible by using up excessive amounts of fat-producing food elements.

i. Injury Susceptibility. Susceptibility to injury is reduced through exercise. Muscles, tendons, and joints are strengthened. Injuries such as hernia, back strain, and joint sprains are less likely to occur if muscles are maintained in proper tone.

Section IV. Leadership Roles

1401. PSYCHOLOGICAL LEADERSHIP

The full development of a Marine's resources is not all physical. To be effective in developing physical readiness, leaders must realize that mind and attitude are also important to success. The more important psychological considerations are to—

a. Promote Understanding of the Value of Physical Readiness. A desire to be physically ready should be created in all Marines. Motivation is increased and Marines take greater interest in their individual physical fitness if they understand the value and benefits of vigorous exercise. When Marines realize their efforts are an investment in their own personal welfare, it should not be difficult to obtain their cooperation. Marines should understand the objectives, the benefits, and the value of each type of exercise activity in their program. They should also understand the relation of physical readiness to survival in combat.

b. Maintain a Positive Approach. Physical readiness training for combat is strenuous and demanding. It is a responsibility of leadership to create an atmosphere where all desire to participate fully. This attitude should be fostered. A negative approach must not be identified with physical readiness training, even with those having difficulty. Only in unusual cases should fear of punishment be the motivating factor behind good performance. For those few who cannot keep up or attempt to malinger, an effective remedial program is essential.

c. Seek Cooperation and Develop Morale. In a program placing maximum physical stress upon individuals, it is necessary to gain their cooperation. Favorable reaction is enhanced by proper planning and organization, challenging requirements, use of competition, and application of a progressive program resulting in physical fitness. As physical fitness grows, morale also grows.

1402. COMMAND AND SUPERVISORY FUNCTIONS

a. Command Functions. Commanders should take the following actions to support physical readiness training:

(1) Lead by personal example.

(2) Instill command interest and indicate to subordinate personnel the importance of this training to the welfare of the organization.

(3) Allot sufficient time for the achievement of objectives and monitor the use of such allotted time. The substitution of other training or routine duties for scheduled physical readiness training is unsound and unwise.

(4) Assign and properly utilize qualified personnel to supervise and conduct physical readiness training. If leaders are not competent, take action to ensure they become competent quickly. Reassign those who do not meet this standard to other duties.

(5) Make necessary facilities and funds available to support a program to develop physical readiness within all personnel.

(6) Measure the physical fitness of individuals and units in order to evaluate progress and to determine if the program is successful.

b. Supervisory Functions. Leaders responsible for planning, conducting, and supervising physical readiness training should take the following actions:

(1) Prepare physical readiness training schedules which apply the principles of physical conditioning and which aim for a particular type of program plan.

(2) Provide for wide participation of as many Marines as possible. All Marines, regardless of position or age, will benefit from regular exercise. In some instances, special efforts are necessary to overcome obstacles to regular and frequent training. Special effort is also necessary to ensure remedial conditioning. Such conditioning should occur for those who are physically substandard and after extended absence due to leave, sickness, injury, and travel.

(3) Prevent waste or unwise use of time allotted for physical readiness training. Time-wasters include unprepared instructors; assignment of one instructor to a group larger than a platoon; progression which does not keep pace with the physical development of the Marines; extreme formality; inadequate equipment or facilities which require waiting turns to exercise; and lengthy rest periods between exercises which interfere with the application of overload.

(4) Ensure that the program contains vigorous physical activity. Such activity places progressively greater demands upon the body during each exercise session and also over the duration of the training program. To be of benefit, exercise must tire the muscles and cause the heart to increase its rate of beat.

(5) Set an overall objective for each physical fitness program.Observe the training as necessary to ensure that the established objectives are being achieved.

(6) Observe physical readiness training to insure the use of a positive approach. To implement a positive attitude, small-unit leaders and instructors should personally set the example; have an understanding, fair, and sympathetic attitude; recognize individual differences; and motivate Marines toward their best effort.

(7) Guide and inform small-unit leaders and instructors concerning approved techniques, directives, and literature. As necessary, arrange for local training of instructors to include clinics, conferences, schools, and demonstrations.

(8) Determine the effectiveness of physical readiness training by personal participation in and observation of training, analysis of field inspection reports, and analysis of individual physical fitness test scores. Scores may be combined to reflect the fitness of the unit.

1403. SMALL-UNIT LEADERS AND INSTRUCTORS

a. Responsibility. The instruction and conduct of physical readiness activities are the function of company/battery commanders, platoonleaders and persons assigned as instructors. Experience has proved the effectiveness of physical fitness development when conducted in company-and platoon-size units under direct control of the leader with overall supervision by the parent-unit commander. For example, all the platoons of a company may exercise at the same time under the general supervision of the company commander, with each platoon conducting the assignment separately and under its own leadership.

b. Leader's Assignment. A small-unit leader or an instructor in a school or training activity is assigned to a combat unit or to a support unit. In this assignment, the leader is responsible for all training to include physical readiness training. In a different situation, the noncommissioned officer (NCO) or officer may beassigned as a full-time physical readiness instructor. This contrasts to a unit leader assignment where only part of the time is devoted to such training. In either case, the leader will hold an important and vital position for the physical fitness of Marines.

c. Leader's Training. Leaders may come to the assignment either fully or partially trained. It may be their first responsibility for the development of physical fitness. If they have had previous training through experience, make certain that their information is supplemented with study of this manual. If they have had professional training in physical education during civilian life, but no military experience, they should also use this manual to learn the methods used by the Marine Corps. A new leader should take advantage of various ways to learn including attendance at leader training courses, self-study, practice, and discussion with more experienced leaders.

d.Leader's Objective. As a physical readiness training instructor, the leader has two general objectives. The first is to motivate Marines to want to be physically fit. The second is to conduct a program that will develop a high degree of physical fitnss. Motivated Marines will react enthusiastically to such a program. It aids greatly in achieving local program objectives.

e. Leader's Personal Fitnes. A unit leader who must instruct and demonstrate physical activities must be in physical condition to do the job without undue physical stress.The leader should be able to do those things that must be demonstrated. The leader's strength, endurance, posture, and skill should set the example. This does not mean that the leader must excel, as other Marines do not expect championship performance. However, they do expect, and deserve, a creditable showing of fitness for the job.

f. Leader's Knowledge. The leader must have three types of knowledge to properly administer physical readiness training. Theyare- -

(1) Knowledge of Marines. The leader must understand Marines, know how to lead and motivate them, understand how they learn, and apply this knowledge wisely in the day-to-day training situation.

(2) Understanding of Body Functioning. A more intelligent exercise program results from understanding and applying the principles which govern physical conditioning of the body. The leader with such knowledge can better prescribe, adjust, and regulate exercise types, amounts, and progression to attain fitness.

(3) Understanding Exercise Activities. The leader needs to understand the contribution each type of physical activity makes to physical fitness, and how to use each activity to develop fitness. Skill to demonstrate and lead the various activities is a necessary part of technique and is invaluable to the instructor or small-unit leader.

Chapter 2

PHYSICAL READINESS TRAINING PROGRAMS

Section I. Development of a Program

2101. GUIDANCE FOR PLANNERS

This chapter instructs planners on physical readiness training procedures. It contains program planning guidance including factors to consider when developing programs; steps in assembling a program; definitions of activity packages and systems of exercise; selection of activity packages; and selection of systems used in implementing those packages.

2102. MARINE CORPS MISSION

The mission of the Marine Corps is to seize and defend advanced naval bases and to perform other missions as the President of the United States may direct. Traditionally, to perform this mission, Marines have been projected into the area of operations by Navy vessels and have attacked hostile beaches over the shore in landing craft, assault amphibian vehicles, and helicopters. Recently, performance of this task has been somewhat complicated by the introduction of the concepts of maritime prepositioning and the air-landed Marine expeditionary-brigade. Under these concepts, Marines, landed at airfields in foreign countries, will join their equipment at commercial ports. The distance between these airfields and ports is often 10 or more miles. In time of international crisis, it can be expected that transportation assets will not be adequate. Marines from all elements of a Marine Air-Ground Task Force may have to march from the airfield to the site of their equipment. All Marines must be physically conditioned to rigors of conducting foot marches with individual weapons and equipment.

2103. EVALUATION OF FITNESS

The Marine Corps mission is such that all Marines must constantly and consistently achieve a high level of fitness that prepares them for the demands of combat. While Marines have many different MOS's, all Marines must be prepared for the demands of marching under load and performing basic infantry tasks, such as rear security and patrolling. While the physical fitness test remains our universal measure of individual fitness, it is the commander's responsibility to observe and evaluate the unit's ability to perform effectively in combat. Often the commander's most effective evaluation tool is weekly physical training sessions. The commander should participate in these sessions and see that they are fully integrated into the unit physical conditioning program.

2104. INTEGRATION OF TRAINING

Training time must be used efficiently and wisely. Every opportunity to integrate physical training into other training activities should be seized. For example, foot marches under load can be integrated into the normal daily activities as a way to move from one training or work site to another. This integration reflects realism in training and should be used to maintain overall proficiency. To accomplish this requires imagination and enthusiasm on the part of the commander. The area of operations in which the training is conducted determines what can be integrated and how. If possible, training should also consider the terrain and climate of the area in which the unit will subsequently conduct training or operations. Training should include familiarization with special equipment and the application of specialized techniques to tactical principles. Activities such as gun drills and command post emplacement drills also have physical conditioning value, particularly if supervised and done competitively or measured against time standards.

2105. ACTIVITIES AT THE UNIT LEVEL

Physical readiness training is a command responsibility and is generally conducted in Marine units at the company and platoon levels. Battalion/squadron commanders must constantly supervise, making sure that all Marines are physically prepared for combat.

a. Physical readiness training should usually be conducted weekly by each battalion/squadron-level unit, using events designed for all personnel (e.g., marches under load, battalion/squadron runs, military field days, etc.).These events enable the commanders to demonstrate personal leadership, and observe and evaluate the physical condition and combat physical readiness of their Marines. This exercise period should be designed to elevate the unit's morale and emphasize unit identity through the wearing of uniforms and the carrying of battalion and company guidons. The battalion/squadron commander should make every effort to be a visible participant in the unit's physical readiness program on a weekly basis. These activities could be conditioning marches under load or a weekly calisthenics/unit run.The suggested times for these activities are early on Monday morning or late on Friday afternoon after the training week has been completed. The uniform may be physical training gear orutilities and boots.

b. Commanders should be cautious of long runs in boots and utilities. This is not to suggest that units should not run in utilities, but to remind commanders that boots are designed for marching, not running. If commanders choose boots and utilities, they should select off-road routes which will provide a cushioning effect.

c. Units should be made well aware of the commander's intent before the unit falls out for exercise. If the exercise period is conducted early in the morning, it need not necessitate an extensive shower/clean-up period afterward, so long as all Marines from the commander on down remain dirty. Commanders should inform their Marines not to fall out for these sessions in clean, pressed uniforms.

d. The physical readiness training of headquarters units is often difficult to manage. Often physical training is left for individuals or sections to develop and manage on their own. This approach can be effective, but if not supervised, it can result in a "paper" program. Headquarters unit commanders should endeavor to conduct weekly integrated training which will enable them to evaluate the physical readiness of their personnel. Often this training has to be scheduled before or after normal working hours to maximize participation of the entire unit.

2106. STEPS IN PLANNING

To implement workable and effective programs (as directed by CMC ALMAR 261/87) the planner must-

a. Determine the Type of Program Needed. Marine units are inherently different in organization and mission. The physical readiness program must be tailored to the mission and to the current physical condition of most unit personnel. Programs to meet this need are of the following types:

(1) **Developmental Programs.** Marines in a beginning or poor state of physical readiness need a program which will develop strength, endurance, physical skills, and character traits which are beneficial to successful accomplishment of military missions. Such programs should be applied progressively to rise gradually to a peak of fitness and skill.

(2) Maintenance Programs. Once Marines reach the sustaining stage of conditioning, their goal is then to maintain this level by participation in a maintenance program.

(3) Remedial Programs. The term "remedial" is usually applied to those individuals or groups who possess substandard physical fitness. For example, a remedial physical conditioning program could be applied to persons who are overweight, who fail to reach physical fitness test standards, or who have missed extended periods of conditioning due to illness, injury, extended hospitalization, or other absence. Saturday training sessions are designed as remedial sessions for all individuals who either miss the daily training or have failed to meet adequate standards. Note that personnel in light duty or no-duty status should be expectedto attend all training sessions, observing, supervising, or assisting as needed.

b. Determine the Time Required. The amount of time for training operations varies considerably. However, every unit can find time to conduct physical readiness training. Frequent (i.e.,daily) physical training of short duration-30 to 60 minutes-is preferred over occasional longer periods. Other demands for training time are so urgent that every minute of time allotted for physical readinesstraining should be used. Determine the time required per week anddivide it into daily blocks.

c. Organize for Various Group Size. It is essential to stress exercise rather than formality. Marines must complete the program where they are- -on the training field, in the motor pool, on the range, next to the classroom, in the office area, in the shop, aboard ship, or elsewhere. Although desirable, it is not always possible to assemble company-size units for physical training. Platoon-size groups are appropriate for the proper conduct of physical conditioning activities. Certain situations may require exercise programs for section- or squad-size units. This manual outlines programs for all situations and types of organizations.

d. Allow for Weather and Exercise Area. In programing and scheduling, the climate and terrain often govern the selection of activities.

(1) Weather changes cause differences in temperature, rainfall, wind chill, and snow. These changes should be anticipated as they dictate the type of program. Alternate plans should be part of the schedule. Nevertheless, normal weather changes and rain should not drive training inside.

(2) Local terrain and available exercise areas may also influence the selection of activities and the type of program which it is possible to support. Some activities can be completed in nearly any area.

e. Plan for Seasonal Change. As most physical readiness training is conducted outdoors, it is necessary to recognize seasonal change. A program should be divided into fall, winter, spring, and summer parts. Seasonal change also causes change in light. For example, an early morning program started in the summer will have ideal lightconditions, yet in the fall or winter, darkness will occur at that same hour and interfere with the conduct of the program. Develop programs in seasonal blocks, and make provisions for anticipated changes.

f. Consider Needed Facilities. An excellent program can be conducted with practically no facilities since there are exercises which require no equipment. However, a better program can be developed when supported by certain facilities and items of equipment. Proper command support, plus ingenuity, will solve this problem. (Items of equipment, when necessary to support the recommended exercises, are included in the chapters on exercise activities.)

g. Specify Appropriate Uniform. The uniform worn for exercising depends upon the season of the year, the state of the weather, and local regulations. Whenever possible, Marines should be dressed alike. Undershirts are preferred as the upper garment when the weather permits. A uniform that restricts the free movement of the body should not be worn when exercising.

h. Consider Availability of Instructors. Leaders who can lead and direct the scheduled activity must be available. Organizational units should train junior officers and noncommissioned officers down to squad or section leaders to instruct and lead the various activities.

i. Select Activity and System. With the type of program needed and the objectives in mind, select an activity package or an integrated training package for each day's scheduled physical training. At this same time, the system to be used in employment of the selected activity or activities must be determined. This selection affects equipment, areas, instructors, transportation, and other support requirements.

j. Secure Command Participation and Support. Prepare and brief the commander to assure full understanding of the objectives and administration of the program. The full participation and support of the commander will greatly improve the level of success.

k. Supervise Execution. Determine needs, publish the program, and supervise its execution as a necessary part of developing the unit's training schedule.

2107. SELECTION OF ACTIVITIES AND SYSTEMS

a. Exercise Activities. The planner must determine exercise activities which will be appropriate to include in the program. An exercise activity is a single means of exercise usually identified by the name applied; for example, running, log exercises, and obstacle course. Many exercise activi ties can become part of activity packages.

b. Activity Packages. Many of the physical activities described in this manual are arranged in prescribed sequences and are known as activity packages. An activity package is a number of exercises of the same type, assembled as a group or a set, and arranged in a specific sequence. Exercise packages are organized in such manner that not more than 15 minutes will be required to complete the execution of any package. Each type of activity is explained in laterchapters. The number of available drills, tables, or circuits; the manner of organization; and the contribution each makes to the total program are covered. Full understanding of this information will greatly assist in developing effective programs. Various designations are used to identify exercise packages; for example, conditioning exercises when arranged in a set order are known as drills, and other packaged activities are designated as tables or circuits. The following are activity packages:

(1) Activity packages for groups:

o Conditioning Drill Two
o Conditioning Drill Three
o Rifle Drill
o Log Drill
o Grass Drill
o Running Tables
o Guerrilla Tables
o Circuit-Interval Table
o Combatives Tables
o Relay Tables

(2) Strength Circuits:

o Fixed Circuit
o Movable Circuit
o Simplified Circuit (Circuit Interval Table)

(3) Activity packages for individuals:

o The 6-12 Plan
o Weight Training
o Isometric Exercise

c. Advantages of Using Packages. These of exercise packages simplifies scheduling and conducting of exercise and results in the following benefits:

(1) Schedule development is simplified as the planner assembles packages which will satisfy the training objective. There is no need to deal with selection of individual activities or to be concerned about the amount of time to be expended on each.

(2) Any 15-minute period, and in some cases less time, can be scheduled or used to perform an activity.

(3) If longer periods of time are available or if the objective demands, several packages can be assembled to provide a more complete period of activity.

(4) Marines are assured a balanced set of exercises or activities as each package is carefully arranged to reach all muscle groups.

(5) The instructor can concentrate on the conduct of a vigorous workout as the type and duration of the activity have already been determined.

d. Nonpackaged Activities. Several types of activity are not packaged. Activities in this category are conditioning marches, unit runs, obstacle courses, combat water survival swimming, teamcontests, and team sports. These activities can be scheduled in combination with packaged activities, or they may be scheduled separately. Often they require a longer period of time. Most nonpackaged activities require a 50-minute period to satisfactorily complete their objective. For conditioning marches, a few hours are required. The benefits of these activities should not be overlooked as some desirable objectives cannot be attained without them.

e. Systems of Exercise. Several methods or systems of organizing exercise and activity packages can be used. Each system is based upon a specific organization as follows:

(1) Single Activity System. The unit leader immediately assumes command of the unit at the beginning of the exercise period. The leader moves the unit to a predesignated exercise site at double time, forms the unit in a circle around the leader, groundsclothing and equipment as appropriate, and quickly moves into the exercise routine. There is usually no time to teach; therefore,the Marines must know the activity to be used. At the conclusion of a 5- to 15-minute period, the unit leader returns the unit to the instructional area at double time and releases the unit for the next scheduled activity.

(2) Progressive Activity System. All Marines (company or platoon) complete activities in the same order during the period. For example, Drill One is followed by dual combatives, and finally a 1-mile run. This system is usually progressive from a warm up activity, such as Drill One, to an activity which contributes in a major way to one of the objectives, such as aggressiveness development through combatives or cardiovascular development, such as running.

(3) Rotating Activity System. Set up the same number of activities or stations as there are platoons in the company. Each platoon rotates through each station in turn. With four platoons in a 50-minute period, about 10 minutes can be devoted to each station. With three platoons, approximately 15 minutes may be spent at each station. Activities must be of a type that can be covered in the time allotted. For example, with three platoons, Station 1 could be Conditioning Drill One; Station 2, running; and Station 3, a team contest.

(4) Circuit System. Set up a number of stations to provide various types of exercise equipment and/or items of apparatus. The idea is to keep all Marines busy and exercising vigorously for a short period of time at each station. The fixed strength circuits and the movable strength circuits are examples of this system. Station changes must be rapid, and the exercise must be started quickly after each change. Since the objective is to exercise at top speed, the motivation comes from frequent change of activity by moving to another station. Rotation by station groups continues until all Marines have covered all stations.

(5) Interval System. This system stresses the development of strength and endurance. It involves heavy work for a given distance within a specified time, alternated with lighter work and recovery, but never stopping during the workout. This procedure is repeated, and the intensity is increased gradually as physical condition improves, but always with adequate recovery. The important concept is stress, recover, stress, recover, and so on. This system is often applied through running, but other activities of a continuous nature may also be used. An example is the Circuit-Interval Table.

f. Steps in Selecting Activities and Systems. The planner should follow these steps.

(1) Evaluate the needs of one unit with emphasis upon unit mission, objective, and time available.

(2) For each day, select an activity package or a combination of activities which will contribute to the objective.

(3) Then determine the system to be used in implementing the selected activities. There are several possible choices and much opportunity for flexibility in program development. For example, the weekly program may contain various exercise packages, systems, and time periods. Additional guidance is contained in following chapters.

2108. SAMPLE WEEKLY PROGRAMS

The variety of weekly programs given shows available alternatives. Programs and schedules other than those illustrated may be assembled to provide schedules to fit any situation. Many other scheduling combinations are possible through use of the packaged activities, varied time periods, and different types of assigned duty.

a. Sample A. Figure 2-1 illustrates a 60-minute time allotment in which the single activity system is used with activity packages that vary from day to day. In this program, 5 hours of training are scheduled for the week. This program needs some type of integrated unit training to supplement it each week. A bimonthly conditioning march of 10 miles in 3 hours is suggested to round out this program. The Saturday session is designed for remedial physical training.

NOTE: A physical training activity must be conducted at a minimum of once every 72 hours. After 72 hours, muscles begin to atrophy.

Figure 2-1 is available in the illustration section at the end of the book

b. Sample B. A more comprehensive weekly program (fig. 2-2) illustrates the scheduling of various-length periods, a variety of activities, and the use of three systems during the week. In this schedule, 3.5 hours are included for the week. The following should be noted concerning this schedule:

Figure 2-2 is available in the illustration section at the end of the book

(1) On Monday, a 60-minute period is available for physical readiness training. Three 20-minute packages are scheduled. Each platoon, under its own leadership, will progress through each scheduled activity in turn.

(2) On Tuesday, Wednesday, and Thursday, only 30-minute periods are available. Here each platoon leader supervises each platoon in a single activity system.

(3) On Friday, the company commander directs training to exhibit personal leadership example and supervision. In this example, only one strength circuit and one cross-country course are available, and each will accommodate only one platoon. For this reason, Conditioning Drill One station is included. A platoon rotates to each station within the 60-minute period.

(4) Thc Saturday session once again is designed for remedial physical training.
(5) Bimonthly, a 10-mile/3-hour march under load will be necessary to round out this program.
NOTE: Cardiovascular/respiratory exercise to improve endurance must be of long enough duration (20 minutes or more) to induce and maintain oxygen debt.

c. Sample C. A third sample schedule (fig. 2-3) illustrates the use of the single activity system for 5.25 hours of training. This schedule includes sustaining-type activities for a unit that has passed through both the toughening and slow improvement stages of conditioning. The use of competitive activities is featured to holdinterest and provide self-motivation. A bimonthly 10-mile/3-hour march under load will be necessary to round out this program. The Saturday session once again is designed forremedial physical training.

Figure 2-3 is available in the illustration section at the end of the book

NOTE: Thc body will respond to hard work or physical training without soreness and stiffness provided the work or physical training is conducted regularly.
NOTE: Excrciscs of short duration increase strength and bulk but do little to effect endurance.
NOTE: Thc times shown in the charts arc participation times.

Section II. Remedial Physical Training

2201. NEED FOR ATTENTION

Remedial physical training is a process by which physically substandard individuals are conditioned to meet standard levels of performance. To achieve a full degree of operational readiness, it is necessary to bring all Marines up to the prescribed standard. Experience has demonstrated that some Marines have difficulty due to a poor state of fitness, obesity, or lack of motivation. Attention to these individual deficiencies will help improve unit combat readiness.

2202. IDENTIFICATION OF INDIVIDUALS

The company or battery commander identifies persons who cannot achieve the prescribed level of proficiency through the physical fitness test or as a result of observation during training. The commander notes particular weakness of body parts as indicated by failure of certain test events. These Marines are then placed in a special remedial program either at platoon, company/battery, or battalion/squadron level. Extra time is devoted to overcoming the weakness. These extra periods of conditioning may be during or after duty hours, as determined by the commanders. Other persons need reconditioning as result of hospitalization. Restoring physical fitness to damaged areas of the body is accomplished through progressively graded physical activities under professional supervision, not by the physical fitness trainers.

a. Types of Deficiencies. Physical deficiencies which can be corrected by exercise fall into several categories.

(1) Lack of Strength in One or More Body Parts. The major muscle areas concerned are the arms and shoulder girdle, back, abdomen, and legs.

(2) Lack of Overall Endurance. Usually, there is deficient muscular and cardiovascular endurance.

(3) Deficiency in Coordination and Agility. In these cases, physical skill is not developed to a satisfactory degree in activities such as crawling, running, jumping, climbing, traversing, vaulting, pushing, pulling, lifting, and carrying.

(4) Overweight or Underweight. Either condition may interfere with physical fitness and mission accomplishment. Lack of exercise is not always the cause. The cause may be malfunction of normal physiological functions or it may be poor health habits such as over- or under-eating, lack of adequate rest, or overconsumption of alcohol.

(5) Lack of Motivation. Not all persons are motivated to attain or maintain **a** desirable state of fitness. Some Marines do not understand the importance of physical fitness, some find proper exercise too difficult, and others find it inconvenient.

b. Causes of Deficiencies. There are several causes for Marines being deficient in physical fitness:

(1) Absence of exercise.

(2) Exercise which fails to develop all muscle groups and components of fitness.

(3) Exercise which is not vigorous enough or which lacks progression.
(4) Injury or illness which depletes fitness.
(5) Inadequate amounts of sleep or rest.

c. Methods of Detection. The commander has several means by which to detect physical deficiencies:

(1)
from those who pass. Make an analysis to determine the cause of failure as revealed by test scores.

(2) Observe Marines as they perform physical tasks-
o Marines who have difficulty during training or in physical types of work,
o Marines who have become obese and therefore experience difficulty.

(3) Be alert to those Marines who are often on sick call, returnees from hospitalization, or newly assigned.

2203. ADMINISTRATION OF REMEDIAL ACTION

a. Group. Attitude Marines who are singled out as being physically deficient are self-conscious and are not always convinced they need extra help. Within a deficient group, motivation may be low. These Marines must be convinced that a remedial program which is tailored to their needs will help them and will eliminate their deficiencies. Encouragement is often needed and desirable.

b. Leadership. The leader of this group must study individuals and know their deficiencies. The unit leader should counsel Marines individually, observe them closely as they progress through the remedial program, maintain records, and adjust the program as required.

c. Measurement.. Whenever improvement in performance is noted, Marines should be measured by the physical fitness test either individually or by group. When testing reveals an individual to be satisfactory, that individual should be released from the remedial group. There may be exceptions to this policy in the case of Marines who are overweight or in the execution of an exercise program prescribed for some special purpose.

d. Organization of a Remedial Group

(1) A remedial group is usually a small group of Marines within a company or battalion. In some situations, it may be a larger group numbering 50 to 75 Marines. In the case of a larger group, divide the Marines initially into subgroups according to ability. Prescribe exercise loads commensurate with their ability. General conditioning is sufficient in the early part of the program to qualify Marines who are on the borderline. Qualification will reduce the size of the group, permitting more individual and specialized attention to those who remain.

(2) As the program progresses, regroup individuals who have like deficiencies in order that they can concentrate on their weakness. For example—

(a) One group may be weak in the arms and shoulders as revealed by failure in the chin up event. This group, with an assistant instruction charge, can work on pull-ups, rope climb, pushups, rifle or logexercises, horizontal ladder, and similar types of exercise.

(b) Another group may be weak in general endurance as revealed by the3-mile run. This group could profit from participation in conditioning drills, running, grass drill, and strength circuit.

(3) It may be necessary to form some groups to overcome weaknesses in skills, such as an inability to throw, to quickly change direction while running, to crawl rapidly, or to carry a load. Lack of coordination or lack of practice may cause these deficiencies. In this instance, instructors must provide an opportunity to practice and correct poor form and other errors as they are noted.

Chapter 3
PHYSICAL CONDITIONING ACTIVITIES
Section I. Marching Under Load

3101. GENERAL

Few physical fitness activities are as directly related to readiness for combat as foot marches under load. In addition to the obvious physical conditioning and unit cohesiveness benefits, marches under load prepare most Marines for numerous foreseeable tasks in combat. The idea that only infantry and reconnaissance units actually require training marches to prepare them for combat is clearly a mistake. In addition to the very real possibility of Marines from all types of units being used to fill combat-depleted infantry units, it is likely in maritime prepositioning force operations that aviation, combat service support, and command element personnel will have to march with weapons and equipment, from the arrival airfield to the equipment marshalling site. In American coastal cities, the harbor is nearly always 10 or more miles from the airport. It is difficult to imagine any different condition in a lesser developed country .A further combat-related benefit is that foot marching under load is the most effective way to develop leg strength in the context of a unit training program. In his classic study of troop performance in World War II, Men Against Fire, S.L.A. Marshall noted that the greatest single weakness of replacements of all occupational fields was leg strength. Truck drivers must push trucks which are stuck in mud, and headquarters communicators must climb hills to emplace antennas. The demands on the lower body in combat apply to all Marines, and foot marching is one activity which can help build these muscles while being conveniently integrated with the rest of the training program.

3102.TRAINING GOALS

The standard for success of a foot march is very simple to measure: did the unit arrive at the destination at the prescribed time with Marines in condition and required equipment present to accomplish the mission?A progressive program can increase a unit's readiness. It can instill pride by increasing distances and rates of march and by selecting routes over increasingly challenging terrain as the Marines become better conditioned. However, the ability to execute the mission at the conclusion of the march must remain the standard of success.

3103.MOVEMENT PLANNING

The basic considerations in planning a foot march are the mission, tactical situation, terrain and weather, and the units to participate. The success of the march will depend largely upon the thoroughness with which it is planned. A successful march is characterized by adherence to prescribed routes and time schedules, the efficient employment of the means available, and the ability of the unit to accomplish its assigned mission upon arrival at the destination.

a. Movement Orders. Movement planning culminates in the preparation and issuance of an operation order prepared in the standard, five-paragraph format. Necessary annexes are attached to furnish detailed information required for the movement. Written movement orders are rarely prepared at company level. A discussion of the annexes is contained in this chapter.

b.March Planning. March planning, as discussed here, is the planning conducted at battalion level. March planning may be organized into the following steps:

(1) **Preparation and Issuance of the Warning Order.** In order to afford subordinate units the maximum possible time to prepare for a pending move, a warning order containing all available information about the march is issued. The amount of planning time available will determine the time of issuance and the content of the warning order.

(2) Estimate of the Situation. In his estimate, the commander considers the mission, terrain, weather, time, and space factors, available routes, available transportation for the movement of equipment and/or shuttling of Marines, enemy capabilities, disposition of own forces, physical condition and training of Marines, and courses of action available to the command.

(3) Organization and Dispatch of a Reconnaissance. Party very march plan is based on as thorough and complete a ground reconnaissance as time and the situation will permit. Map and aerial reconnaissance are valuable in formulating a plan, but are not a substitute for ground reconnaissance. Route reconnaissance is accomplished by a reconnaissance party which usually consists of a reconnaissance element, an engineer element from the attached or supporting engineer unit, and a traffic control element. Unit standing operating procedures generally establish the basic composition of the reconnaissance party. It is modified as necessary to meet the requirements of a particular march. A recommended method of reporting information obtained by the route reconnaissance party is shown in figure 3-1.The minimum information required from the reconnaissance party is—

(a) Available routes and conditions.(Routes may be specified by higher headquarters.)

(b) Recommended rate of march.

(c) Selection of start point and release point, or confirmation of the suitability of start point and release point previously selected by map reconnaissance.

(d) Confirmation of location of the assembly or bivouac area.

(e) Location of critical points on the route.

(f) Distance between critical points on the route and total distance from start point to release point.

Figure 3-1 is available in the illustration section at the end of the book

(g) Location of obstacles and estimation of necessary Marines and equip- ment needed to repair and maintain routes.

(h) Number of guides required and their location on the route.

(4) Development of Detailed Movement Plan. Some of the elements included in a detailed movement plan are-

(a) Organizing the Column. To facilitate control and scheduling, units will be organized into serials and march units and given an order of march. In determining the order of march, the march planner must consider the enemy situation and the desirable order of arrival of the units at the destination. Where dispersion is required, a unit may be organized into two or more columns, each assigned a different route.

(b) Using Reconnaissance Information. Results of the route reconnaissance will be used to select the route(s); determine the start point, critical points along the route(s), and the release point; and select the rate of march.

(c) Determining March Computations. March unit pass time is based on the strength, formation, and rate of march. The pass time of the marching columns, plus necessary time distance computations, will be used to determine the completion time of the march.

(d) Drafting of Road Movement Table. Using the completed march computations, a draft road movement tabic is compiled.

(e) Checking the Plan. Using the draft road movement table and a road movement graph, the movement plan is checked to ensure that it conforms to the directive of the higher headquarters and the battalion commander's instructions.

(5) Preparation and Issuance of the Road Movement Order. After the plan has been checked and approved by the commander, an operation order is prepared and issued. The operation order may be in written form or issued orally, and is accompanied by a road movement table, overlay and/or strip map, and appropriate administrative details.

(6) Road Movement Table. A road movement table, prepared as an annex to the operation order, provides serial commanders with arrival and clearance times at critical points along the route of march. It also provides the column commander with information as to the proposed location of elements of the column at various times.

(7) Overlay/Strip Map. An overlay serves the normal purpose and should show, as a minimum, the present location of units, route of march, critical points, and the new location of units at the destination. A strip map is a schematic diagram of the route of march and shows landmarks and critical points with the distances between them. A strip map may be issued as an annex to the road movement order, in addition to or in lieu of an overlay.

c. March Computations. Prior to issuance of the operation order, the S-3 must verify time and space computations as they provide him with the necessary data for the preparation of a road movement table.

(1) Time-Distance. Time distance (TD) is determined by dividing the distance to be traveled (D) by the rate of inarch (R):

TD = D/RWhere TD = hours D = distance in kilometers R = kilometers per hour

(2) Length of ColumnThe length of column (LC) is used to determine the pass time of a column. The sum total of the following two parts determines the length of column. (See pars, (a) and (b)).The two parts are the space occupied by the Marines alone (including the distance between Marines) and the sum of the distances between the units of the column (column gap).

(a) The length of column of Marines alone is determined by multiplying the number of Marines by the appropriate factor selected from the table below. The length of column does not include distances between units. LC Marines = No. of Marines x factor.

(b) The total distance in meters between units is determined as follows:

Step One. Determine the number of gaps between serials (total serials minus one).

Step Two. Multiply the number of serial gaps from Step One by the length (in meters) between respective units.

Step Three. Determine the number of gaps between march units (total march units minus one, minus the number of serial distances).

Step Four. Multiply the number of march unit gaps from Step Three by the length (in meters) between respective units.

Step Five. Add the totals from Step Two and Step Four to get the total meters for the column gap.

EXAMPLE:A battalion foot column is organized into 12 platoon-sized march units and 3 company-sized serials. REQUIRED: total column gap distances when there are 100 meters between serials and 50 meters between march units.

(3)Pass-Time.Pass-time (PT) is the time a unit takes to pass a specified point. For foot columns, the pass-time is determined by applying the following formula:PT (minutes) = LC x FACTOR (for appropriate rate of march).

PASS-TIME FACTORS - FOOT TROOPS

.0150 for 4.0 km/h
.0187 for 3.2 km/h
.0250 for 2.4 km/h
.0375 for 1.6 km/h

EXAMPLE: Determine the pass-time of a unit whose length of column is 1,500meters and is marching at a rate of 4 km/h.PT (min) = 1,500 x .0150 (the factor for 4.0 km/h) = 22.5 mm.

(4) Completion Time. Completion time is the time of day that a march will be completed. Completion time is determined by using the following formula: Completion Time = SP time + TD + PT + Scheduled Halts (other than normal hourly halts).

EXAMPLE: A column's starting point (SP) time is 0700.The time-distance is6 hours and 40 minutes. Pass-time of the column is 30 minutes. A 35-minutelunch halt has been scheduled. What is the completion time of the march?Employing the 24-hour clock system, the formula can be applied as follows for simplified addition of the times:

HrMin SP time................0700
TD......................0640
PT......................0030
Lunch halt.............0035

Completion time.........13 105or 1445

The march will be completed atl445.

(5) Experience Tables. Based on previous movements made by a unit, data is accumulated to facilitate march planning. Such data includes approximate pass-times lor various elements of the battalion. The S-3 can utilize these data rather than computing them each time a march is scheduled. Such experience tables serve to reduce the time required to complete the computation phase of march planning. Matter appropriate to the unit standing operating procedure should be integrated therein.

d. Road Movement Graph

(1) A road movement graph is a time-distance diagram used in planning, preparing, or checking road movement tables, and for controlling marches. The graph provides a visual representation of a march plan so that conflicts and discrepancies may be prevented in the planning stage, before congestion occurs on the route. It is not usually issued as a part of the order. Road movement graphs may be applied to small units, to movements of a single column, or to a large organization scheduling separate elements, moving by various means, with different rates of march, over one or more routes.

(2) To construct a road movement graph, use the following steps: (See fig. 3-2.)

Figure 3-2 is available in the illustration section at the end of the book

Step One. Determine the number of hours available for the march. Designate the lower left corner of a sheet of graph paper as the SP time or the earliest even hour before the march is to begin. Select a scale and plot the hours available in sequence from the left on the horizontal axis.
Step Two. Determine the distance to be moved in kilometers. Indicate the SP at the lower left corner of the graph sheet and, using an appropriate scale, plot the number of kilometers on the vertical scale from the SP to the release point. Indicate the location of critical points on the vertical scale.
Step Three. At the proper distance from the SP, draw a horizontal line indicating the location of the RP. Indicate by a vertical line the hour when the movement must be completed. Plot lines representing route restriction, if any, at the proper distances and times on the graph.
Step Four. Determine the pass-times of foot and motor elements of the column.
Step Five. Starting at the SP at the specified hour, plot the movement of the head of the leading element. If the rate of march is 3 kilometers per hour, the lead element will have moved 1.5 kilometers in 30 minutes, 3 kilometers in 1 hour, etc. Plot the trace of the lead man to the RP. The last Marine will cross the SP in pass-times after the lead man. Measure this time on the graph and plot the trace of the last Marine of the column. The lines describing the head and tail of a march serial are parallel. Indicate the time subsequent serials reach the SP and plot the trace of the head and tail. Check to see that the plan complies with all restrictions and orders. If there are any violations or conflicts, the plan may be made to conform by changing the order of march, the starting time of the serials, the planned rate of march, or the organization of the column.

e. Conversion to the Metric System. The adoption of the metric system for expressing linear distances requires a simple means of conversion.
(1) To accomplish this, use the following tables:
o Multiply inches x 2.54 to obtain centimeters. Multiply centimeters x .39 to obtain inches,
o Multiply yards x .91 to obtain meters. Multiply meters x 1.1 to obtain yards,
o Multiply miles x 1.6 (or 8/5) to obtain kilometers. Multiply kilometers x .62 (or 5/8) to obtain miles.

(2) Linear distances can be accurately measured on maps which have been converted to the metric system by using a straightedge graduated in centimeters (cm).The centimeter scale can be used in lieu of the graphic scale when the scale of the map is shown, as follows:

Map Scale:	One cm Equals:
1:500,000	5,000 meters
1:250,000	2,500 meters
1:100,000	1,000 meters
1:50,000	500 meters
1:25,000	250 meters

3104. INDIVIDUAL LOAD

Since the backbone of Marine ground combat forces is the infantry, tactical mobility is largely dependent on the marching ability of the individual Marine. The load that each Marine must carry is the greatest impediment to mobility. In general, the commander must make every effort to reduce the individual load of his Marines to the absolute minimum. While the load has the greatest effect on Marines marching on foot, the value of carrying the minimum essential supplies and equipment applies for all Marine forces.

a. Fighting Load. The fighting load consists of items of clothing, equipment, weapons, and ammunition that are carried by, and essential to, the effectiveness of the combat Marine and the accomplishment of the immediate mission. The fighting load should not weigh more than 40 pounds. The commander must ensure that his Marines' loads- -and his own- -be stripped to the bare minimum. The addition of small, light-weight items in excess of the limit should not be tolerated, since cumulatively, these items will take a toll in energy. Every extra pound a Marine must carry decreases his combat effectiveness.

b. Existence Load. The existence load consists of items other than those in the fighting load that are required to sustain or protect the combat Marine, that may be necessary for increased personal and environmental protection, and that are not normally carried by the individual. The existence load is normally carried by the Marine's unit. Marching, but not engaged in combat, a Marine may be expected to carry a maximum of 50 pounds of supplies and equipment.

c. Considerations

(1) The primary consideration is not how much a Marine can carry, but how much he can carry without impaired combat effectiveness-moral or physical. The combat strength of a unit is not counted simply in numbers of Marines, but in the numbers of willing and physically able Marines.

(2) It is generally better to risk temporary inconvenience from lack of health and comfort items than to exhaust Marines due to overloading.

(3) It is a fundamental truth that men become physically exhausted more quickly when under the stress of combat. Marines must be required to carry less into battle than they are conditioned to carry in training. Marines should be conditioned for carrying weight, but should be equipped in combat for fleetness of foot.

(4) A common mistake is to base the fighting load on the gear and supplies necessary to meet every contingency. The commander cannot reasonably expect the Marines to carry enough gear for every possible eventuality. The items to be carried must be based on reasonable expectations.

(5) It is the commander's responsibility to procure the transport to carry any additional gear. As a rule of thumb, a rifle company, or unit of similar size, requires one 5-ton truck and trailer in support to carry additional gear. In cold weather, or other conditions in which the necessary personal gear increases, this requirement will increase.

(6) The commander must ensure that the supply system provides, on a dependable and timely basis, the balance of essential supplies and equipment not carried by the unit. Marines must feel confident that they will be supported with the necessary supplies and equipment.

(7) In training, commanders must implant in their Marines the pride in operating under austere conditions. To effectively demonstrate the need for this spartan mentality, the commander must personally set the standard. Marines should be trained in field craft, foraging techniques, and the use of caches and field expedients. Maximum use should be made of captured stores.

3105. MARCH TECHNIQUES AND PROCEDURES

a. Organization. A command executing a march is organized into march units, march serials, and march columns, as necessary for control.

(1) March Units. A march unit is a unit of command which moves and halts at the command of a single commander. The march unit normally corresponds to one of the smaller Marine units such as a platoon or company.

(2) March Serials. A march serial consists of one or more march units organized under the senior officer and given a numerical or alphabetical designation to facilitate control. A serial is usually a battalion, but may be a company if the battalion is marching alone.

(3) March Columns. A march column is composed of elements of a command moving over the same route. It may be composed of one or more serials. To facilitate control, a column commander is designated. A column is normally a regiment or larger unit, but may be a battalion if marching alone.

b. Order of March. The order of march is determined by the mission, terrain, and the commander's desired order of commitment of units into action.

c. Control and Coordination Measures. The commander establishes initial control of the march by designating control measures in his march order. The most frequently used measures include:

o Start point and release point.
o Time at which head or tail of column passes the start point.
o Rate of march.
o Order of march.
o Route of march.
o Location of command post.
o March communications.

d. The Pacesetter. The pace-setter is an experienced individual carrying the same load as the majority of the Marines and marching from 4 to 10 meters at the head of the column. (See fig. 3-3.)The pace-setter's primary duty is to maintain the rate of march ordered by the column commander.

(l) Setting the Pace. The pacesetter does this by establishing the pace (length of step) and cadence (steps per minute) to obtain the prescribed rate of march. He should be of medium height so normal strides will be taken. Overstriding or understriding tends to tire leg muscles quickly and affects the combat efficiency of marching Marines. The officer marching at the head of the column supervises the pacesetter to ensure that he takes normal strides and maintains a uniform cadence.

Figure 3-3 is available in the illustration section at the end of the book

(2) Reducing Accordion Effect. Accordion effect occurs when the column alternately lengthens and shortens, causing the rear elements of the column to double time to maintain distance. It is caused by changes in the rate of march by the lead elements, usually after passing a slowing obstacle. To reduce this effect as much as possible, lead elements should slow their rate of march for a sufficient time after crossing obstacles to allow the rear elements to maintain the prescribed distance without running. The rear elements can signal the head when the last man has cleared the obstacle and the rate can be resumed. Since some accordion effect is nearly inevitable, the order of march should be periodically rotated to prevent units in the rear from becoming physically exhausted at the conclusion of the march.

e. Length, Rate and Pace

(1) **Length of March.** The length of march varies depending on terrain and weather, enemy situation, and physical and mental condition of Marines. The normal length of march for a 24-hour period is from 20 to 32 kilometers (12 to 20 miles) marching from 5 to 8 hours at a rate of 4 kilometers (2.5 miles) per hour (km/hmi/h).A march in excess of 32 kilometers in a 24-hour period is considered a forced march. Well-trained units that have been progressively trained will be able to achieve a unit goal of 60 miles in a 3-day period.

(2) Rate of March. The same factors of terrain, weather, load to be carried, and condition of Marines affect the rate of march. The commander considers these and selects a rate which will place his unit at the destination in the shortest possible time in condition to accomplish the mission. Unit standing operating procedure should state the standard rate of march for that unit over normal terrain for both road and cross-country movement and in both day and night. The commander then modifies this rate if the situation requires. Normally prescribed standards are as follows:

Figure is available in the illustration section at the end of the book

(3) Pace and Cadence. The normal pace is 30 inches. A pace of 30 inches and a cadence of 106 steps per minute result in a speed of 4.8 kilometers per hour or 3 miles per hour and a rate of 4 kilometers per hour or 2.5 miles per hour if a 10-minute rest halt per hour is taken. Since the pace of each man may vary, the cadence may have to be adjusted to maintain the prescribed rate. Slope of ground and surface footing affect pace. A normal cadence is relatively simple to maintain on level or gently rolling terrain on firm footing. Mud, sand, loose gravel, and hills all greatly affect pace.

f. Halts. Halts during a day's march are taken at regular intervals to rest personnel and adjust loads .Halts are specified by standing operating procedures or march order. Such factors as midday heat or enemy action may force the adoption of long midday halts or night marches. At long halts, each unit disperses to a previously selected location near the route of march.

(1) Normal Time and Frequency. Under normal conditions, a 15-minute halt is made after the first 45 minutes of marching. Following the first halt, a 10-minute halt is made after each 50 minutes of marching. Variations of this schedule are made when a scheduled halt time occurs when passing through a built-up area or when cover and concealment are required by the tactical situation and none is available. Observation posts may be established, if required for security of the unit during halts.

(2) Actions at Halts. All units in the column should be halted at the same time. At the halt signal, Marines should move to the side of the road, staying within the immediate vicinity of their unit. Marines should remove or loosen gear and sit or lie with feet elevated. Commanders inspect Marines and equipment, and corpsmen administer medical treatment, as required.

g. Communications. The four primary means of communications used in controlling foot marches are radio, visual, sound, and messenger.

(1) Radio. Radios are normally used for communications between platoons and higher headquarters in the march column. Radio transmissions should be held to the absolute minimum necessary for control and should be at minimum power required.

(2) Visual.. Visual communications may include flashlights, luminous markers, panel sets, pyrotechnics, and hand and arm signals. When contact with the enemy is not expected, the loss of light discipline from some of these signals may be a lesser security risk than the risk of disclosure from radio transmissions.

(3) Sound. Sound communications include voice and such signaling devices as horns and whistles. Some of their uses include assembling Marines at the conclusions of halts and providing warnings of air or chemical attacks.

(4) Messengers. Messengers are particularly effective during periods of radio silence and reduced visibility. Messages should be simple and brief to preclude mistakes by the messenger.

h. Security. Both active and passive measures are used against attack by enemy aircraft and indirect fire weapons during movement. Active measures include the use of air sentries and organic and attached weapons in accordance with the unit air defense standing operating procedure. Passive measures include the use of concealed routes and assembly areas, night marches, and extended interval between elements of the column. With imagination and planning, all of these measures can be practiced in training without detracting from the physical conditioning benefit of the march. In fact, such drills as rapidly taking cover during simulated attacks and then quickly resuming the march can prove physically challenging.

i. Reconnaissance. Reconnaissance determines in advance critical points along the march route such as bridges, fords, and obstacles in order that congestion or delay may be held to a minimum and local security provided. Paragraph 3103.C describes organization of the reconnaissance party.

3106.TACTICAL MARCHES

These marches are movements of Marines and equipment not in direct ground contact with the enemy but expecting early ground contact either en route or upon arrival at the objective. Some characteristics of this type of movement include security elements to the front, flanks, and, if necessary, to the rear of the march column. Additional considerations include the selection of concealed areas on defendable terrain for start and release points and rest stops. In training, infantry units of regimental size or smaller will normally derive the greatest benefit from tactical marches. The security requirements usually dictate a slower rate of march and, therefore, reduce the physical conditioning value for other types of units which would seldom move in combat with the expectation of early ground contact with the enemy. Two key points should be remembered regarding tactical marches in a training program.

a. Security. Although security requirements often slow the rate of march, it is a valid and important training objective for infantry units to work toward reducing march times while still practicing sound tactical security. March standing operating procedures, good land navigation skills, and simple but complete march orders contribute as much to this goal as physical conditioning. The idea that proceeding slowly enhances security is seldom true. Keeping the enemy off balance by rapid movement frequently is the best guarantee of security.

b. Training. In peacetime training, it is critical that support personnel who normally are located with infantry units in combat, participate completely with infantry units in forced march training. Not only must they not slow the infantry rate of march because of inferior conditioning, but they must be thoroughly familiar with the infantry unit's standing operating procedure for security on the march. Personnel from within the ground combat element such as artillery forward observer teams and combat engineer units, can be easily identified for this training and should be habitually associated with the same infantry units. Personnel from other elements, such as low-altitude air defense teams and radio battalion detachments, are more difficult to identify and earmark for support of the same infantry unit. Their normal mode of employment is in vehicles and their units frequently provide general support vice direct support. This does not, however, lessen their requirement to be prepared to provide support to dismounted infantry over terrain or in a tactical situation not conducive to vehicles. Such training needs to take place, and planning and cooperation among the elements of the MAGTF can overcome the obstacles.

3107.ADMINISTRATIVE MARCHES

These marches are movements of Marines and equipment made when no enemy interference is expected except from aircraft or indirect fire. During these marches, units are administratively grouped for ease of control and speed of movement. The term administrative foot marches should not be taken to imply that tactical considerations are disregarded. There are simply less stringent security measures used than when ground contact with the enemy is a distinct possibility. The normal formation is the route column with one file moving on either side of the road and with negligible security. Air sentries are used and dispersion between individuals and units is practiced. Although roads usually provide the fastest route, there may be reasons to conduct these marches cross-country (particularly as part of a training program). Administrative marches are the type of foot marches most practical for training programs for non-infantry units. These marches may also play a key role in the physical conditioning program of infantry units, because of the greater speeds possible due to lessened security requirements.

3108. FORCED MARCHES

A forced march is a march which covers a greater distance than 32 kilometer (20 miles) in a 24-hour period. Normally the extra required distance should be achieved by increasing the number of hours marched in a day beyond the norm (8 hours) rather than increasing the rate of march beyond the norm (4 kilometers per hour/2.5 miles per hour). However, there will be occasions when the situation or mission demands an increase in the rate of march.

a. Maximum Recommended Distances. The maximum recommended distances for forced marches are-56 kilometers (35 miles) in 24 hours; 96 kilometers (60 miles) in 48 hours; or 128 kilometers (80 miles) in 72 hours.

b. Sample Time Breakdown. A sample time breakdown for a forced march of 52 kilometers, beginning at first light is as follows:

Figure is available in the illustration section at the end of the book

3109. NIGHT MARCHES

Night marches are characterized by closed formations, more difficult control and reconnaissance, and a slower rate of march than day marches.

a. Control. Control is increased by reducing the distance between individuals and units, and by using connecting files to maintain contact between platoons and companies. Connecting files are normally constituted from personnel from the rear most march unit.

b. Safety on Roads. Because of reduced visibility, night marches on roads used by vehicles require attention to safety procedures to prevent accidents. Commanders should not think these safety measures apply only to peacetime training. In combat, nearly all vehicles operate with blackout lights at night, making visibility even more difficult than in garrison driving. If consistent with the tactical situation, the following measures can reduce the possibility of marchers being struck by vehicles:

(1) Use off-road trails and routes as much as possible.

(2) Guards to the front and rear of march units should be marked with strips of luminous tape and carry red-filtered flashlights. These measures are generally consistent with light discipline, while still affording reasonable warning to approaching drivers.

Section II. Conditioning Drills (One, Two, and Three)

3201. GENERAL

a. Description and Objective. Conditioning drills are calisthenic exercises. Each drill contains seven exercises organized and numbered in a set pattern. Each drill takes 15 minutes to complete. The objective of conditioning drills is to exercise all major muscle areas in order to develop strength, endurance, coordination, and flexibility.

b. Area and Equipment. Any level area is satisfactory for conduct of the drills. Drills One and Two contain ground exercises. If ground conditions are unsatisfactory, Drill Three should be used as it contains no ground positions. Usually, no equipment is required;however, if the group exceeds a platoon in size, an instructor's stand is necessary.

c. Formation. The extended rectangular formation is used in this drill. (See app. A.)

d. Starting Level and Progression. The starting level is six repetitions of each exercise. An increase of one repetition for each three periods of exercise in which the drill is performed is an acceptable rate of progression. This rate is continued until 12 repetitions can be completed. To maintain, continue the drill at 12 repetitions. To progress, move to 6 repetitions at a more difficult drill which exercises the same muscle group. Progression can also be gained by moving from Drill One to Drill Two, as Drill Two is more demanding.

e. Starting Positions. Starting positions vary with the exercise and are explained in each exercise.

f. Leadership. A principal instructor demonstrates and leads the drills. The instructor must be familiar with leadership techniques peculiar to conditioning drills to include the exercises, commands, counting cadence, cumulative count, formation, method of teaching the exercises, and utilization of assistant leaders.

g. Use With Other Programs. Conditioning Drills One, Two, and Three reach all major muscles of the body. They are easy to learn and to perform, and they are simple to administer and supervise. These features, coupled with the short time required for completion, the fact that no equipment is necessary, and adaptability to most areas of execution, make these drills possible in any programs.

3202. CONDITIONING DRILL ONE

This conditioning drill is similar to the calisthenics drill commonly known throughout the Marine Corps as the "Daily Seven."The "Daily Seven" may be used as a substitute for Conditioning Drill One.

a. Exercise l:High Jumper

(1) Starting Position. Feet separated shoulder width, knees flexed, body bent forward at the waist, arms aligned with the trunk and hips, elbows locked, palms facing, fingers extended and joined, head and eyes to the front.(See fig. 3-4, A.)(Elbows remain locked throughout the exercise.)

Figure 3-4 is available in the illustration section at the end of the book.

(2) Cadence. Moderate.

(3) Movement. A four-count exercise: at the count of—

(a) ONE -- Take a slight jump into the air, swinging the arms forward and up to shoulder level.

(b) TWO -- Take a slight jump into the air and swing the arms downward and back, returning to the starting position.

(c) THREE -- Take a vigorous leap into the air, swinging the arms forward and up to an overhead position, momentarily looking skyward, on returning to the ground the knees are flexed, head and eyes return to the front.

(d) FOUR--Repeat the action of count two.

b. Exercise 2: Bend and Reach

(1) Starting Position. Feet spread more than shoulder width, arms overhead, elbows locked, palms facing, fingers extended and joined, head and eyes to the front.(See fig. 3-4, B.)

(2) Cadence. Moderate.

(3) Movement. A four-count exercise: at the count of—

(a) ONE-Bend at the knees and waist, swing the arms straight downward and reach between the legs. Touch the ground as far to the rear as possible and look to the rear. (Elbows remain locked throughout the exercise).

(b) TWO-- Recover sharply to the starting position.

(c) THREE--Repeat the action of count ONE.

(d) FOUR-- Repeat the action of count TWO.

c. Exercise 3: Pushup

(1) Starting Position. Front leaning rest position. To assume this position there is a silent one-two count:on the silent count of one, assume the squatting position, heels together, elbows locked inside the knees, hands flat on the ground directly beneath the shoulders; on the silent count of two, thrust the legs to the rear, toes and heels together, body straight from head to heels.(Seefig. 3-4, C.)

(2) Cadence. Moderate.

(3) Movement. A four-count exercise: at the count of—

(a) ONE -- Flex the elbows lowering the body until the thick portion of chest touches the ground.

(b) TWO -- Raise the body until elbows are straight and locked.

(c) THREE-- Repeat the action of count ONE.

(d) FOUR -- Repeat the action of count TWO. (On returning to position of attention, the silent one-two count is used in reverse).

d. Exercise 4: Trunk Twister

(1) Starting Position. Feet are spread more than shoulder width apart, fingers placed behind neck, thumbs pointing downward, elbows back. (See fig. 3-4, D.)(Elbows remain well back throughout the exercise).

(2) Cadence. Slow.

(3) Movement. A four-count exercise: at the count of-

(a) ONE-- Keeping the knees locked and the back straight, bend forward at the waist sharply, with a slight recovery.

(b) TWO-- Twist the trunk to the left vigorously at the waist, keeping the elbow back. The left elbow is higher than the right.

(c) THREE-- Twist vigorously to the right, so the left elbow comes under the right.

(d) FOUR-- Straighten sharply to the starting position. NOTE: Do not attempt to touch the elbows to the knees on counts two and three.

e. Exercise 5: Squat Bender

(1) Starting Position. Feet are spread less than shoulder width apart, hands on hips, thumbs in small of back, elbows back. (See fig. 3-4, E.)

(2) Cadence. Moderate.

(3) Movement. A four-count exercise: at the count of—

(a) ONE-- Assume the squatting position, maintain balance on the balls of the feet, with trunk erect thrust arms forward to shoulder level, elbows locked, palms down.

(b) TWO-- Recover to starting position. Elbows are well back.

(c) THREE-- Keeping the knees locked, bend forward at the waist, touching the ground in front of the toes.

(d) FOUR-- Vigorously recover to the starting position.

f. Exercise 6: Body Twist

(1) Starting Position. On the back, arms extended sideward on the ground, palms down. The legs are raised to a near vertical position, feet together, knees locked.(See fig. 3-4, F.)

(2) Cadence. Slow-fast.

(3) Movement. A four-count exercise: at the count of-

(a) ONE-- Lower legs slowly to the left until they touch the ground near the left hand, keeping the knees straight and shoulders on the ground.

(b) TWO-- Recover the starting position by quickly raising the legs, keep knees straight and feet together.

(c) THREE-- Repeat movement of count ONE, except the movement is to the right side.

(d) FOUR-- Recover sharply to the starting position.

g. Exercise 7: Stationary Run

(1) Starting Position. Position of attention. (See fig, 3-4, G.)

(2) Cadence. Fast.

(3) Movement.

(a) At the command of execution, start running in place at double-time cadence, lifting the left foot first time cadence. Follow the instructor as he counts two repetitions of cadence;

e.g., 1, 2, 3, 4--1, 2, 3, 4. The instructor then gives informal commands such as FOLLOW ME, running on the toes and balls of the feet, keeping the back straight, speeding up the cadence to a sprint, raising the knees high, leaning forward at the waist, and pumping the arms vigorously.

(b) To halt the exercise, the instructor will count two repetitions of cadence as the left foot strikes the ground: 1, 2, 3, 4 - 1 , 2, 3, HALT.

NOTE: When counting cadence, the instructor counts only as the left foot strikes the ground. The duration of the exercise is approximately 1 1/2 minutes.

3203. CONDITIONING DRILL TWO

a. Exercise 1: Jumping Jack

(1) Starting Position. Feet separated more than shoulder width, arms overhead.(See fig.3-5, A.)

(2) Cadence. Moderate.

(3) Movement. A four-count exercise: at the count of~

(a) ONE-- Jump to position with the feet together and assume the squatting position, swinging the arms sideward and downward, placing the hands palms down on the ground, elbows locked inside the knees.

(b) TWO-- Recover to the starting position by jumping to the side straddle and swinging the arms sideward overhead.

(c) THREE-- Repeat the action of count ONE.

(d) FOUR-- Recover to the starting position.

b. Exercise 2: Turn and Bend

(1) Starting Position. Side straddle, arms overhead.(See fig. 3-5, B.)

(2) Cadence. Moderate.

(3) Movement. A four-count exercise: at the count of-

(a) ONE-- Turn the trunk to the left and bend forward over the left thigh, attempting to touch the fingertips to the ground outside the left foot. Keep the left knee straight. On successive repetitions attempt to touch farther and farther to the side.

(b) TWO-- Recover to the starting position.

(c) THREE-- Turn the trunk to the right and bend forward over the right thigh, trying to touch the hands to the ground outside the right foot. Keep the right knee straight.

Figure 3-5 is available in the illustration section at the end of the book.

(d) FOUR-- Recover to the starting position.

c. Exercise 3: Eight-Count Pushup

(1) Starting Position. Position of attention. (See fig. 3-5, C.)

(2) Cadence. Moderate.

(3) Movement. An eight-count exercise: at the count of—

(a) ONE-- Assume the squatting position, palms on the ground directly beneath the shoulders, elbows locked inside the knees.

(b) TWO-- Thrust the legs to the rear assuming the front leaning rest position.

(c) **THREE**-- Flex the elbows until the thick portion of the chest touches the ground.

(d) **FOUR**-- Raise the body on a straight plane until the elbows are locked.

(e) **FIVE**-- Repeat the action of count THREE.

(f) **SIX**-- Repeat the action of count FOUR.

(g) **SEVEN**-- Recover to the squatting position as in count ONE (elbows locked inside the knees).

(h) **EIGHT**—Return sharply to the position of attention.

d. Exercise 4: Turn and Bounce

(1) Starting Position. Feet spread more than shoulder width apart, arms sideward at shoulder level, palms up. (See fig. 3-5, D.)

(2) Cadence. Slow.

(3) Movement. An eight-count exercise: at the count of—

(a) **ONE**-- Turn sharply to the left as far as possible, then recover slightly.

(b) **TWO**-- Again turn to the left as far as possible and recover as in ONE.

(c) **THREE**-- Repeat the action of count TWO.

(d) **FOUR**-- Recover sharply to the starting position.

(e) **FIVE**—Turn sharply to the right as far as possible, then recover slightly.

(f) **SIX**—Again turn to the right as far as possible and recover as in FIVE.

(g) **SEVEN**-Repeat the action of count SIX.

(h) **EIGHT**-Return to the starting position.

NOTE: The head and hips remain to the front throughout the exercise and the knees and elbows are locked at all times.

e. Exercise 5: Squat Stretch

(1) Starting Position. Attention. (See fig. 3-5, E.)

(2) Cadence. Moderate.

(3) Movement. A four-count exercise: at the count of—

(a) **ONE**-- Squat, placing the hands on the ground about 12 inches in front of the feet.

(b) **TWO**— Keeping the fingertips on the ground, straighten the knees completely and raise the hips.

(c) **THREE**-- Recover to position ONE.

(d) **FOUR**-- Recover to the starting position.

f. Exercise 6: Leg Circular

(1) Starting Position. On the back, arms stretched sideward, palms down, feet raised foot from ground, knees straight. (See fig 3-5,F.)

(2) Cadence. Slow.

(3) Movement A four-count exercise: at the count of—

(a) ONE-- Swing the legs as far as possible to the left, keeping the knees straight and the legs together.

(b) TWO-- Swing the extended legs over head with the thighs as close as possible to the trunk.

(c) THREE-- Swing the legs as far as possible to the right.

(d) FOUR-- Recover to the starting position.

g. Exercise 7: Stationary Run

(1) Starting Position. Position of attention. (See fig. 3-5, G.)

(2) Cadence. Fast.

(3) Movement

(a) At the command of execution, start running in place at double-time cadence, lifting the left foot first. Follow the instructor as he counts two repetitions of cadence; for example: 1, 2, 3, 4--1, 2, 3, 4. The instructor then gives informal commands such as FOLLOW ME. Running on the toes and balls of the feet, keeping the back straight, speeding up the cadence to a sprint, raising the knees high, leaning forward at the waist, and pumping the arm vigorously.

(b) To halt the exercise, the instructor will count two repetitions of cadence as the foot strikes the ground: 1, 2, 3, 4-1,2, 3, HALT.

NOTE: When counting cadence, the instructor counts only as the left foot strikes the ground. The duration of the exercise is approximately 1 1/2 minutes.

3204. CONDITIONING DRILL THREE

a. Exercise 1: Side Straddle Hop

(1) Starting Position. Position of attention. (See fig. 3-6, A.)

(2) Cadence. Moderate.

(3) Movement. A four-count exercise: at count of-

(a) ONE-- Take a slight jump into the air, moving the legs sideward (more than shoulder width apart); at the same time, swing the arms overhead (to an overhead position) clapping the palms together.

Figure 3-6 is available in the illustration section at the end of the book.

(b) TWO-- Take a slight jump into the air, swing the arms sideward and downward returning to the starting position.

(c) THREE-- Repeat the action of count ONE.

(d) FOUR-- Repeat the action of count TWO.

b. Exercise 2: Back Bender

(1) Starting Position. Standing, feet 12 inches apart, fingers placed behind the head. (See fig. 3-6, B.)

(2) Cadence. Slow.

(3) Movement. A four-count exercise: at the count of-

(a) ONE-- Bend the upper trunk backward, raising the chest high, pulling the elbows back, and looking upward. Keep the knees straight.

(b) TWO-- Recover to the starting position.

(c) THREE-- Repeat the action of count ONE.

(d) FOUR-- Recover to the starting position.

c. Exercise 3: Squat Thrust

(1) Starting Position. Position of attention. (See fig. 3-6, C.)

(2) Cadence Moderate.

(3) Movement. A four-count exercise: at the count of—

(a) ONE-- Assume the squatting position; heels together, placing the hands flat on the ground, shoulder width apart, elbows locked and inside the knees.

(b) TWO-- Thrust the legs to the rear, assuming the front leaning rest position, body in line from head to toe, heels and toes together.

(c) THREE-- Return to the squatting position as in ONE.

(d) FOUR-- Return to position of attention.

d. Exercise 4: Side Bender

(1) Starting Position. Feet are spread more than shoulder width apart, arms are raised sideward and overhead, thumbs interlocked palms to front, fingers extended and joined, elbows locked. (See fig. 3-6, D.)

(2) Cadence. Slow.

(3) Movement. An eight-count exercise: at the count of—

(a) ONE-- Bend to left as far as possible, then recover slightly.

(b) TWO-- Again bend to the left as far possible, then recover slightly.

(c) THREE-Repeat the action of count TWO.

(d) FOUR-- Recover sharply to the starting position.

(e) FIVE-- Bend to the right as far as possible, then recover slightly.

(f) SIX-- Again bend to the right as far as possible, then recover slightly.

(g) SEVEN-- Repeat the action of count SIX.

(h) EIGHT-- Recover sharply to the starting position.

NOTE:Keep the elbows and knees locked throughout the exercise. The bend should occur to the side and not the front.

e. Exercise 5: Knee Bender

(1) Starting Position. Feet are spread less than shoulder-width apart, hands on hips, thumbs in small of back, elbows back. (See fig 3-6, E.)

(2) Cadence. Moderate.

(3) Movement. A four-count exercise: on the count of—

(a) ONE-Do a knee bend, lean trunk forward at the waist, thrust arms between legs until the extended fingers touch the ground palms to the ground, hands 6 inches apart.

(b) TWO~Recover sharply to the starting position.

(c) THREE-Repeat the action of count ONE.

(d) FOUR-Repeat the action of count TWO.

f. Exercise 6: Bottoms Up

(1) Starting Position. Front leaning rest position. A silent one-two count is used as in the pushups. (See fig. 3-6, F.)

(2) Cadence. Moderate.

(3) Movement A four-count exercise: at the count of-

(a) ONE-- With the weight on the hands, and knees locked, jump forward bringing the feet as close to the hands as possible; look to the rear.

(b) TWO--Keeping the knees locked, thrust the legs backward assuming the front leaning rest position.

(c) THREE-- Repeat the action of count ONE.

(d) FOUR-- Repeat the action of count TWO.

g. Exercise 7: Stationary Run

(1) Starting Position. Position of attention. (See fig. 3-6, G.)

(2) Cadence. Fast.

(3) Movement

(a) At the command of execution, start running in place at double time, lifting the left foot first. Follow the instructor as he counts two repetitions of cadence; for example: 1, 2, 3, 4--1, 2,3, 4. The instructor then gives informal commands such as FOLLOW ME, running on the toes and balls of the feet, keeping the back straight, speeding up the cadence to a sprint, raising the knees high, leaning forward at the waist, and pumping the arms vigorously.

(b) To halt the exercise, the instructor will count two repetitions of cadence as the left foot strikes the ground: 1, 2, 3, 4 - 1, 2, 3, HALT.

NOTE: When counting cadence, the instructor counts only as the left foot strikes the ground. The duration of the exercise is approximately 1 1/2 minutes.

Section III. Rifle and Log Drills

3301. RIFLE DRILL

a. Description and Objective. Rifle exercises are conditioning exercises performed with a rifle. Each drill contains six exercises (fig. 3-7) and they are numbered in a set pattern. The drill takes 15 minutes to complete. The objective of rifle drills is to exercise the arms, shoulders, and back muscles in order to develop strength and endurance, particularly in the upper body. In units without rifles, log drills may be substituted.

b. Area and Equipment. Any level area is satisfactory for conducting this drill. Each Marine completes these exercises from a standing position and no ground contact is required. Each Marine will need a rifle and, if the group exceeds a platoon in size, then the instructor will need an instructor's stand.

c. Formation. The extended rectangular formation is used in this drill. (See app. A.)
d. Starting Positions. Starting positions vary with the exercises and are explained in each exercise. As in all set conditioning drills, the command used to start the exercise is STARTING POSITION, MOVE. The following directions apply to rifle drill.
(1) In those exercises which start from the rifle downward position, on the command MOVE, execute port arms as prescribed in NAVMC 2691, Drill and Ceremonies Manual, and then assume the starting position. The command to return to the men to the position of attention at the conclusion of the exercise is POSITION OF ATTENTION, MOVE.
(2) In exercises which terminate in the rifle downward position, on the command MOVE, execute the position followed by order arms as prescribed in NAVMC 2691.
(3) In the exercises which terminate in a position other than the rifle downward position, Marines first assume the rifle downward position before executing port arms and order arms.
(4) These movements are executed without command. This procedure promotes uniformity, but precision is not expected. To be effective, rifle exercises must be strenuous enough to tire the arms, but not to the point where the arms cannot move with precision.

e. Leadership. A principal instructor demonstrates and leads the drill. He must be familiar with leadership techniques for conditioning exercises and the peculiar techniques for rifle drill.

3302. EXERCISES PERFORMED WITH RIFLES

The exercises of rifle drill are outlined in the following paragraphs,
a. Exercise 1: Foreup, Behind Back

(1) Starting Position. Rifle downward, feet together.(See fig. 3-7, A.)

Figure 3-7 is available in the illustration section at the end of the book.

(3) Movement. A four-count exercise: at the count of—

(a) ONE-- Swing the arms forward and upward to the overhead position. Inhale.

(b) **TWO**-- Lower the rifle to the back of the shoulders. Exhale.

(c) **THREE**-- Recover to position ONE and inhale.

(d) **FOUR**-- Recover to the starting position and exhale.

b. Exercise 2: Lunge Side, Turn and Bend

(1) Starting Position. Rifle downward, feet together.(See fig. 3-7, B.)

(2) Cadence. Moderate.

(3) Movement. An eight-count exercise: at the count of

(a) **ONE**-- Lunge sidewards to the left, swing the rifle forward and upward to the overhead position.

(b) **TWO**-- Turn the trunk to the left and bend forward over the left hip. At the same time, swing the rifle to a **low** horizontal in front of the left ankle.

(c) **THREE**-- Recover to position ONE.

(d) **FOUR**-- Recover to the starting position.

(e) **FIVE, SIX, SEVEN, and EIGHT**-- Repeat on the right side.

c. Exercise 3: Foreup, Back Bend

(1) Starting Position. Rifle downward, feet together.(See fig. 3-7, C.)

(2) Cadence. Moderate.

(3) Movement. A four-count exercise:at the count of—

(a) **ONE**-- Swing the arms forward and upward to the overhead position.

(b) **TWO**-- Bend backward, emphasizing the bend in the upper back. The face is up.Keep the knees straight.

(c) **THREE**-- Recover to position ONE.

(d) **FOUR**-- Recover to the starting position.

d. Exercise 4:Up and Forward

(1) Starting PositionRifle downward, feet together.(See fig. 3-7, D.)

(2) Cadence. Fast.

(3) Movement. A four-count exercise:at the count of—

(a) **ONE**-Swing the arms forward and upward to the overhead position.

(b) **TWO**- Swing the arms forward to shoulder level.

(c) **THREE**-Recover to position ONE.

(d) **FOUR**-Recover to the starting position.

e. Exercise 5: Foreup, Full Squat

(1) Starting Position. Rifle downward, feet in narrow stance. (See **fig.** 3-7, E.)

(2) Cadence. Moderate.

(3) Movement. A four-count exercise: at the count **of**—

(a) **ONE**-- Swing the arms forward and upward to the overhead position.

(b) **TWO**-- Swing the arms down to shoulder level and assume the squatting position.

(c) **THREE**-- Recover to position ONE.

(d) **FOUR**-- Recover to the starting position.

f. Exercise 6: Arms Forward, Side Bend

(1) **Starting Position.** Sidestraddle, regular stance, rifle forward.(See fig. 3-7, F.)
(2) **Cadence.** Moderate.
(3) **Movement.** A four-count exercise:at the count of-

(a) **ONE--** Bend the trunk to the left. Keep the knees straight.
(b) **TWO--** Recover to the starting position.
(c) **THREE--** Bend the trunk to the right. Keep the knees straight.
(d) **FOUR--** Recover to the starting position.

NOTE:Keep the rifle on the same level as the shoulders throughout the exercise.

3303. LOG DRILL

a. Description and Objective. Log exercises are conditioning exercises performed with a log. Each drill contains six exercises and they are numbered in a set pattern. The drill takes 15 minutes to complete. The objective of log drills is to develop strength and muscular endurance and, in this instance, under maximum loads. Log exercises also develop teamwork. Log may be used in lieu of conditioning drills after the Marines have become somewhat conditioned.

b. Area and Equipment. Any level area is satisfactory for conducting this drill. Each Marine completes these exercises from a standing position and no ground contact is required. If the group exceeds a platoon in size, then the instructor will need instructor's stand. Each six-person group or eight-person group will need a log. The logs should be from 6 to 8 inches in diameter. They may vary in length from 14 feet (for 6 people) to 18 feet (for 8 people).The logs should be skinned, smoother, and dried. The 14-foot logs should weigh approximately 300 pounds and the 18-foot logs, approximately 400 pounds. Rings should be painted on the logs to indicate each person's position. When not in use, the logs should be stored on a rack to keep them off the ground.

c. Formation. All the Marines assigned to the same log team should be about the same height at the shoulders. The recommended method of dividing the platoon is to have the Marines form a single file or column with short people to the front and tall people to the rear. Have the Marine assume their positions in the column according to shoulder height, not head height. When they are in position, they are given the command COUNT OFF BY SIXES (OR EIGHTS), COUNT OFF, to divide them into six- or eight-person log teams. Each team in turn, can then proceed to the log rack, shoulder a log, and carry it to the designated exercise area. The log teams form in columns in front of the instructor. With the Marine holding the log in the chest position, have them face the instructor and ground the log at least 10 yards between columns and 10 yards between log teams within the columns.

d. Starting Positions. The Marines fall in, facing the log, with their toes about 4 inches from it. The basic starting positions and commands are as follows:(See fig. 3-8.)
(1) **RIGHT-HAND STARTING POSITION, MOVE.** It the command MOVE, move the left foot 12 inches to the left, and lower the body into a flat foot squat. Keep the back straight, head up, and arms between the legs. Encircle the far side of the log with the left hand. Place the right hand underneath the log.(See fig. 3-8, A.)

(2) LEFT-HANDED STARTING POSITION, MOVE. These commands are executed in the same manner as in paragraph a. except that the left hand is underneath the log and the right hand encircles its far side. (See fig. 3-8, B.)

Figure 3-8 is available in the illustration section at the end of the book.

(3) RIGHT SHOULDER POSITION, MOVE. At the command MOVE, pull the log upward in one continuous motion to the right shoulder. At the same time, move the left foot to the rear and stand up, facing left. Balance the log on the right shoulder with both hands.(See fig. 3-8, C.)This movement cannot be performed from the left-hand starting position because of the position of the hands.

(4) LEFT SHOULDER POSITION, MOVE. These commands should be given from the left hand starting position. At the command MOVE, pull the log upward in one continuous motion, to the left shoulder. At the same time, move the right foot to the rear and stand facing right. Balance the log on the left shoulder with both hands.(See fig. 3-8, D.)This movement cannot be performed from the right-hand starting position.

(5) WAIST POSITION, MOVE. From the right hand position pull the log waist high. Keep the arms straight and fingers laced underneath the log. The body is inclined slightly to the rear and the chest is lifted and arched. (See fig. 3-8, E.)

(6) CHEST POSITION, MOVE. This command should be given after the waist position has been assumed .On the command MOVE, shift the log to a position high on the chest, bring the left arm under the log and hold the log in the bend of the arms.(See fig. 3-8, F.)Keep the upper arms parallel to the ground.

(7) To move the log from the right shoulder to the left shoulder, the command is: LEFT SHOULDER POSITION, MOVE. On the command MOVE, push the log overhead and lower it to the opposite shoulder.

(8) To return the log to the ground from any of the above positions, the command is: STARTING POSITION, MOVE. At the command MOVE, slowly lower the log to the ground. The hands and fingers must be kept from under the log.

e. Leadership. A principal instructor demonstrates and leads the drill. He must be familiar with the leadership techniques for conditioning exercises and the peculiar techniques for log drill.

3304. EXERCISE PERFORMED WITH LOGS

The exercises of log drill are outlined in the following paragraphs. Figure 3-9 graphically explains log drill.

a. Exercise l:Two-Arm Pushup

(1) Starting Position. Right or left shoulder position. Regular stance. (See fig. 3-9, A.)

(2) Cadence. Moderate.

(3) Movement. A four-count exercise:at the count of—

- **(a) ONE--** Push the log overhead until the elbows lock.
- **(b) TWO--** Lower the log to the opposite shoulder.
- **(c) THREE--** Repeat the action of count ONE.
- **(d) FOUR--** Recover to the starting position.

b. Exercise 2: Forward Bender

(1) Starting Position. Chest position. Regular stance.(See fig. 3-9, B.)

(2) Cadence. Moderate.

(3) Movement. A four-count exercise:at the count of-

(a) ONE-- Bend forward at the waist, keeping the back and legs straight.

(b) TWO-- Recover to the starting position.

(c) THREE-- Repeat the action of count ONE.

(d) FOUR-- Recover to the starting position.

c. Exercise 3: Straddle Jump

(1) Starting Position. Right or left shoulder position, feet together, fingers interlaced on top of the log.(See fig. 3-9, **C.)**

(2) Cadence. Moderate.

(3) Movement. A four-count exercise:at the count of—

(a) ONE-- Jump to a side-straddle. Pull down on the log with both hands to keep it from bouncing on the shoulder.

(b) TWO-- Recover to the starting position.

(c) THREE-- Repeat the action of count ONE.

(d) FOUR-- Recover to the starting position.

Figure 3-9 is available in the illustration section at the end of the book.

d. Exercise 4: Side Bender

(1) Starting Position. Right shoulder position, feet regular stance.(See fig. 3-9, D.)

(2) Cadence. Moderate.

(3) Movement. A four-count exercise:at the count of—

(a) ONE-- Bend sideward to the left as far as possible, bending the left knee.

(b) TWO-- Recover to the starting position.

(c) THREE-- Repeat the action of count ONE.

(d) FOUR-- Recover to the starting position.

(4) After completing the required number of repetitions, change shoulders and execute an equal number of repetitions to the other side.

e. Exercise 5: Knee Bend

(1) Starting Position. Right or left shoulder position. Narrow stance. Fingers interlocked on top of the log.(See fig. 3-9, E.)

(2) Cadence. Slow.

(3) Movement. A four-count exercise:at the count of—

(a) ONE-- Flex the knees to a quarter-squat position.

(b) TWO-- Flex the knees to a half-squat position.

(c) THREE-- Lower the body to a three-quarter squat position.(Lean slightly forward.)

(d) FOUR-- Recover to the starting position.

NOTE:Pull forward and downward on the log throughout the exercise.

f. Exercise 6: Overhead Toss

(1) Starting Position. Right or left shoulder position, regular stance. The knees are bent to a quarter-squat.(See fig. 3-9, F.)

(2) Cadence. Moderate.

(3) Movement. A four-count exercisc:at the count of-

(a) ONE-Straighten the knees and toss the log into the air approximately 12 inches overhead. Catch the log with both hands and lower it toward the opposite shoulder. As the log is caught, lower the body into a quarter-squat.

(b) TWO-- Again toss the log into the air and when caught return it to the original shoulder.

(c) THREE-- Repeat the action of count ONE.

Section IV. Grass Drills

3401. GENERAL

a. Description and Objective. Grass drills are extremely strenuous exercises and are performed at top speed for only short periods of time. No cadence is counted but the Marines continue to execute the multiple repetitions of the command until the next command is given. The grass drill consists of two drills:Drill One and Drill Two. Each drill contains six exercises. The objective of the drills is to decrease reaction time, to develop cardiovascular endurance, and to provide a vigorous workout for all major muscles.

b. Area and Equipment. Any level area suitable for ground contact and of a size to accommodate the group is adequate. No equipment is needed.

c. Formation. All movements are executed in place. The extended, rectangular formation is recommended for a platoon- or company-size unit. The circle formation is suitable for groups of squad or section size. At the beginning of an exercise program, 2 to 3 minutes of grass drills will insure a good workout.

d. Progression. Progression is gained by gradually increasing the length of time devoted to the drills. As the physical condition of the Marines improves, the periods should be gradually lengthened to 5 minutes. As the second drill is more difficult than the first, some progression can be attained by initially executing Drill One; then as the program and the Marines progress, introduce Drill Two. To extend the duration of the drill, it may be necessary to repeat the drill.

e. Starting Position

(1) The drills are started from the GO position. Other basic positions are FRONT, BACK, and STOP. (See fig. 3-10, A.)

(a) GO. Running in place (top speed): on the toes and balls of feet, knees raised high, arms pumping, body bent forward at waist.

(b) FRONT. Prone position: elbows bent (along body), palms flat on ground directly under the shoulders, legs together and straight.

(c) BACK. Supine position (flat on back): arms extended near side on ground with palms down, legs together and straight, feet toward the stand or instructor.

(d) STOP. Football lineman stance: feet spread and staggered, left arm across left thigh, right arm straight, knuckles on ground, head up, back parallel with ground.

(2) To assume the FRONT or BACK position from the STANDING, GO, or STOP position, vigorously get into the prescribed position as quickly as possible. (See fig. 3-10, B.)

(3) To change from the FRONT to the BACK position, quickly do a pushup, move the feet several short steps to the right or left, lift the arm on the side toward which the feet move, and thrust the legs vigorously to the front. (See fig. 3-10, C.)

Figure 3-10 is available in the illustration section at the end of the book.

(4) To move from the BACK to the FRONT position, sit up quickly, place both hands on the ground to the right or the left of the legs. Move the feet several short steps to the rear on the side opposite the hands. When the feet are opposite the hands, thrust the legs vigorously to the rear and lower the body to the ground.(See fig. 3-10, D.)

f. Leadership. A warm up activity of lesser intensity should proceed grass drill. During the instructional phase and conduct of these drills, the following points should be applied.

(1) The instructor executes only GO and STOP with the Marines. This allows the instructor to supervise the drill.

(2) The commands for grass drills are given in rapid succession without the usual preparatory command.

(3) To prevent confusion, the instructor should give the commands sharply to distinguish them from comments or encouragement.

(4) As soon as the Marines know the drill, they should respond to the instructor's commands and perform all exercises vigorously and as rapidly as possible. All exercises are executed continuously until the next command is given. Insist on top speed performance; anything less is not effective.

(5) The commands peculiar to each exercise are identical to the name of the exercise.

(6) Marines are not to be required to assume the position of attention once the drills are started. To halt the drill for instructions or for rest, the command UP is used. At this command, the Marines assume a relaxed standing position. Do not demand formality. At the conclusion of a fast and vigorous 5-minute grassdrill, it is physically impossible for people to stand at attention.

(7) The sequence of commands for the execution of grass drills should occur as follows:Drill One. GO, FRONT, Bouncing Ball; GO, BACK, Bicycle; GO, Full Squatter; GO, BACK, Situps; GO, FRONT, Mountain Climber; GO, FRONT, Roll Left; GO, STOP, UP.

g. Use With Other Programs. Since grass drills can be executed in a short period of time, they may be executed where only a few minutes are available for exercise or in conjunction with another type of activity. Grass drills are an excellent substitute for running when time is a factor.

3402. GRASS DRILL ONE AND TWO

a.Grass Drill One

(1) Bouncing Ball. From the FRONT position, push up, supporting the body on the hands (shoulder-width apart) and feet. (See fig. 3-11.)Keep the back and legs in line and the knees straight. Bounce up and down by a series of short, upward springs from the hands, hips, and feet simultaneously,

(2) Bicycle. From the BACK position, raise the legs and hips. Keep the elbows on the ground and support the hips with the hands. Move the legs vigorously as if pedaling a bicycle.

(3) Full Squatter. From the STOP position, assume a full knee bend, the feet on line, hands on hips. Bounce up and down in place by short, bouncing jumps.

Figure 3-11 is available in the illustration section at the end of the book.

(4) Situps. From the BACK position and with arms stretched overhead, sit up, reach forward, and touch toes. Return to the supine position.

(5) Mountain Climber. From the STOP position, place both hands on the ground directly under the shoulders. Thrust the right leg to the rear, knee straight. The left foot should be close to the left hand, the left knee outside the left arm. Shift the weight to the hands, thrust off with the rear (right) foot and bring that foot up close to the right hand, the right knee outside the right arm. At the same time, thrust the left leg vigorously to the rear, knee straight. Continue at a fast cadence, alternating the legs.

(6) Roll Left. From the BACK or FRONT position, make one complete roll in the direction commanded. On completing the roll, return to the FRONT or BACK position.

b. Grass Drill Two

(1) Legs Over. .From the BACK position and with arms stretched overhead, palms up, raise the legs upward and then swing them backward over the head until the toes touch the ground behind the head. Return legs to the starting position. (See fig. 3-12.)

(2) V-Up and Touch Toes. .From the BACK position, raise the legs with the knees straight, sit up until the trunk and legs form a V, and touch the toes with the hands. Return to the BACK position.

(3) Rocker. .In the FRONT position, clasp the hands behind the back, arch the body, holding the head back. Start rocking, using the front part of the trunk as a rocker.

(4) Bounce and **Clap Hands**. The procedure is the same as for bouncing ball, but while in the air, clap the hands. This requires a more vigorous bounce or spring. (See fig. 3-12.)

(5) Leg Spreader. .From the BACK position, raise the legs so that the heels are 10 to 12 inches from the ground, spread them apart as far as possible, then close them together. Open and close legs as rapidly as possible.

(6) Forward Roll. .For forward roll from the STOP position, place both hands on the ground, tuck the head, and do one complete forward roll, keeping the legs tucked as you roll, and come back to the STOP position.

Figure 3-12 is available in the illustration section at the end of the book.

Section V.Guerrilla Excrcise

3501.GENERAL

a. Description and Objective. Guerrilla exercises are individual exercises of various basic skills that are performed rapidly while moving forward in a circle formation. There are two tables of guerrilla exercises. Each table takes 15 minutes to complete. The objective of this exercise is to increase strength and endurance, aid flexibility, and develop coordination.

b. Area and Equipment. Any level area is suitable for conducting guerrilla exercises. No ground contact is required other than the hands. There is no equipment requirement.

c. Formation

(1) The circle formation (app. A) is used for guerrilla exercises. Each platoon forms its own circle and engages in guerrilla exercises under a platoon instructor. If the platoon exceeds 30 Marines, double or concentric circles may be used.

(2) When the circle is formed, the instructor steps into the center of the circle and moves clockwise in a small circle. He commands:QUICK TIME, MARCH, 1-2-3-4. (Rapid cadence of approximately 130 counts per minute. Cadence and step are maintained between exercises.)

(3) To reform the platoon after completing guerrilla exercises, the instructor halts the Marines and places the base man or platoon guide where he wishes and commands:

(a) BASE MAN (or platoon guide),POST.

(b) FALL OUT AND FALL IN ON THE BASE MAN (or platoon guide).

d. Progression. Progression may be attained by moving from table I to table II or by shortening the quick time marching periods between exercises and performing all exercises a second time.

e. Leadership

(1) To execute the exercises, the Marines continue at quick time while the instructor simultaneously explains and demonstrates the exercise to be performed, and then commands the Marines accordingly. In each instance, the preparatory command will be the name of the exercise and, in all instances, the command of execution will ***MARCH***.. *To* terminate each exercise, the command *is* ***QUICK TIME, MARCH.*** *The* Marines immediately pick up the step as the instructor counts cadence.

(2) Unless specified differently, each exercise should be continued for 20 to 40 seconds depending upon the vigor of the exercise. The leader can determine the duration of each exercise by observing its effect upon the Marines.

(3) To form for double guerrillas, the commands for pairing the Marines (who are in circle formation) are-

(a) PLATOON, HALT.

(b) FROM (designate an individual), BY TWO'S, COUNT OFF(.Examplel-2; 1-2; 1-2; etc.)

(c) EVEN NUMBERS MOVE UP BEHIND ODD NUMBERS(At this time, adjust pairs according to height and weight.)

(d) YOU ARE NOW PAIRED UP FOR DOUBLE GUERRILLAS. To change the Marines position, merely command CHANGE.)
(e) FORWARD, MARCH.

f. Place in the Program. Many Marines have not had the opportunity to perform the simple skills involved in guerrilla exercises. The conduct of these exercises is a simple matter since they can be performed easily and quickly in almost any situation. The tables of exercise are applicable to all personnel. The tables can constitute a station within a 1-hour period or be completed within a separate 15-minute period.

3502.GUERRILLA TABLES

a. Table I

(1) Double Time.(See fig. 3-13, A.)Hold arms in the thrust position. Execute a double-time run, maintaining the circle formation and the prescribed distance between your fellow Marines. Duration-1 minute.

(2) All Fours.(See fig. 3-13, B.) Face downward. Support the body with the hands and feet. Walk forward hands first.

(3) Crab Walk.(See fig. 3-13, C.) Get in the sitting position, face upward, and lift the hips. Support the body with the hands and feet. Walk forward feet first.

(4) Squat Walk.(See fig. 3-13, D.) Assume a full knee bend position. Grasp the ankles (left ankle with the left hand, right ankle with the right hand). Walk forward.

(5)Broad Jump.(See fig. 3-13, E.) Jump forward on both feet in a series of broad jumps. Swing the arms vigorously to assist the jumps.

(6)Toe-Touch Walk.(See fig. 3-13. F.) Walk forward, bending at the waist and touching one hand to the toe of the opposite foot while it is on the ground. Raise the trunk to the vertical position between steps. Keep the knees straight.

(7)Bottoms-Up Walk.(See fig. 3-13, G.) Assume the front leaning rest position and move the feet toward the hands in short steps, keeping the knees locked. When the feet are as close to the hands as possible, walk forward on the hands to the front leaning rest position.

(8) Straddle Run.(See fig. 3-13, H.) Run forward, leaping to the right from the left foot and to the left from the right foot.

(9) Fireman's Carry .(See fig. 3-13,1.)

(lO)Single-Shoulder Carry(See fig. 3-13, J.) Two men execute the carries as indicated by the diagram. No. 1 man executes one type; No. 2 man executes the other.

Figure 3-13 is available in the illustration section at the end of the book.

b. Table II

(1) Double Time.(See fig. 3-14, A.)Hold arms at the thrust position. Execute a double-time run, maintaining the circle formation and the prescribed distance between your fellow Marine.Duration-1 minute.

(2) Toe-Grasp Walk.(See fig. 3-14, B.) Bend forward and grasp toes. With knees slightly bent, walk forward.

(3) Hand-Kick Walk..(See fig. 3-14, C.) Walk forward, kicking the moving foot upward on every step. At the same time, lean forward and touch the elevated toe with the hand of the opposite arm.

(4) Pike Juinping.(See fig. 3-14, D.) Jump forward and upward from both feet, keeping the knees straight, and at the same time, swing the legs forward and touch the toes with the hands at the top of each jump.

(5) Squat Jump.(See fig. 3-14, D.) Leap forward from the squatting position with the hands on the ground and the arms between the legs. Land on the ground with hands and legs extended. Bring up the legs to the squatting position.

(6) Steam Engine..(See fig. 3-14, F.) Lace the fingers behind the neck and walk forward in the following manner: as the left leg moves forward, raise the knee high, bend the trunk forward, and touch the outside of the right elbow to the outside of the knee. Then lower the left leg and step forward on the left foot and raise the right leg. Repeat with the right leg and left elbow.

(7) Knee-Touch Walk.(Scc fig. 3-14, G.) Walk forward, bending the knees and touching the ground on each step. The knees are bent and straightened on each step.

(8)Hobble Hopping.(See fig. 3-14, H.) Hold foot behind back with opposite hand and hop forward. On the command CHARGE, grasp the opposite foot with opposite hand and hop **forward.**

(9) Cross Carry.(See fig. 3-14,1.)(10)Saddle Back Carry. (See fig. 3-14, J.) Two Marines execute the carries as indicated in the diagram. No. 1 Marine executes onetype;No. 2 Marine executes the other.

Figure 3-14 is available in the illustration section at the end of the book.

Section VI. Running and Orienteering

3601. RUNNING

a. General

(l)Description. The general form and technique for all types of running are fairly constant. (See fig. 3-15.)The head is erect, body slightly forward without bending at the waist, and the arms are at a loose thrust position alternating from front to rear in straight planes. A cross-body arm movement wastes energy .The movement of the legs and feet will be discussed in subsequent paragraphs dealing with the different types of running. Of primary importance is the fact that in all types of running, the toes should be pointed straight ahead. Toeing out is a common error in both running and walking and should be an item of individual correction.

(2) Objective. The objective of running is to develop cardiovascular endurance. Despite the fact that Marines have developed their muscle structure and the strength of their muscle tissue, unless they have developed cardiovascular endurance to a satisfactory degree, they are not entirely physically fit or combat ready. Cardiovascular endurance (wind) depends on the efficiency of the lungs and heart. The efficiency of the lungs and heart depend on the amount of oxygen the lungs can absorb with each breath inhaled and the amount of carbon dioxide the lungs can expel. The process of absorbing oxygen and expelling carbon dioxide (cardiopulmonary process) is performed by the blood that circulates through the lungs. The condition of this process will determine the amount of effort a person can exert over a period of time. Running is one of the best activities to develop this vitally important endurance.

b. Running Skills. In the development of running skills, individuals may require instruction to improve their proficiency. Some important skills to consider are-

Figure 3-15 is available in the illustration section at the end of the book.

(1) Action of the Arms. Arm action is important. Check to see that arms are held loosely and that the action is relaxed. The faster the run, the more rapid the arm action.

(2) Breathing. Allow the individual to breathe through the mouth as the body demands a large supply of oxygen. Oxygen can be inhaled in greater quantities through the mouth.

3602. TYPES OF RUNNING

a. Double Time

(1) Description and Objective. Double timing is marching at the rate of 180 steps per minute, each step being 36 inches in length. It takes practice to double time with precision in formation. The Marines should keep in step, placing their feet flat on the ground. This, however, should not be a stamping motion, but should be done with as slight a jolt as possible. Double timing is like a jog, the difference being that in a jog the feet are lifted well off the ground and the running motion is bouncy. In double timing, the feet skim the ground and there is no bounce to the ran. Double timing is a vehicle for teaching proper running form and for the development of the cardiovascular system.

(2) Area and Equipment. This type of running can be completed over a variety of surfaces. Usually a Marine uses a field or road. There is no equipment requirement.

(3)Progression. There is no set standard for alternating quick time and double time in the early conditioning of Marines. A general rule is to begin with enough quick-time marching to ensure a thorough warming up, then double time about 100 paces. Change again to quick time until the individuals have made a reasonable recovery from the running, then double time another 100 paces. The amount of double time can be increased and the quick time decreased from week to week, until the individuals are double timing about 1800 yards. This type of training should be given at least twice a week, but by no means is it adequate as the sole means of conditioning.

(4) Leadership

(a) The instructor should be to one side of the column or group and toward the rear so there is a full view of all Marines. Inexperienced instructors have a tendency to supervise from a position too far forward.

(b) Select an individual who can maintain the proper cadence to act as the guide during double-time running.

(c) There are several ways for the instructor and group to count cadence while double timing. If not contrary to local policy, learn several methods and use them for variety.

b. Wind Sprints

(1) Description and Objective. This type of running involves a series of 30- or 40-yard dashes, usually conducted in successive waves of squads. Each squad is in line and the squad leader on the right flank. Wind sprints assist indeveloping speed and cardiovascular endurance. Any flat and level area may be used which will permit the squad to form a line and run the required distance.

(2) Progression. One or two 30-yard sprints will be adequate at the beginning. As time passes, sprints can be lengthened and up to six or seven sprints may be used.

(3) Leadership

(a) At the command READY (given by squad leader), each runner assumes the sprinter's starting position. At the command GO, the squad sprints approximately 30 yards, takes 10 yards to stop, and lines up immediately with the squad leader who repeats, READY, GO, and again the squad sprints. At the conclusion of the third sprint, the squad waits until all the squads of the platoon have made three sprints. Then they all line up and the squad leaders conduct three more wind sprints in the opposite direction.

(b) Valuable time is gained by having each squad ready to go when the preceding squad has moved off its second sprint mark.

c. Cross-Country Running

(1) Description and Objective. Cross-country running is a distance run conducted on a course laid out along roads, across fields, over hills, through woods, and on any irregular ground. The cross-country ran may be utilized as a conditioner or as a competitive event; the objective is to cover the distance in the shortest possible time. The course should be 2 to 2.5 miles in length and be laid out to avoid heavy vehicular traffic. The course should be marked by directional arrows until the runners know the course. These runs build leg muscles, increase lung capacity, and develop endurance. Any local area of varied terrain is suitable.

(2) Progression. In the mass training of a large group, leaders should be stationed at the head and the rear of the column and should make every effort to keep the individuals together .After determining the abilities of the individuals in cross-country running, it is advisable to divide the unit into three groups. The poorest conditioned group starts first, and the best conditioned group, last. The starting time of the groups should be staggered so that all of them finish about the same time. In preliminary training, the running is similar to ordinary road work in that it begins with rather slow jogging, alternating with walking. The speed and distance of the run is gradually increased. As the condition of the individual improves, occasional sprints may be introduced. At first the distance run is from 1/2 to 1 mile. It is gradually increased to 2 or 2.5 miles. Well-conditioned personnel can run 2 to 2.5 miles within a 15-minute period.

(3) Leadership. Marines should not be required to take part in distance running until they have been through a progressively scheduled training program which requires a considerable amount of running. Cross-country runs should be scheduled occasionally to provide variety in the program. Cross-country running has the advantage of allowing mass participation. Interest can be stimulated by putting the runs on a competitive basis. As a single activity, short cross-country runs can be scheduled once a week, gradually increasing the distance as the physical conditioning improves.

(4)Use With Other Programs. Cross-country running can be combined with other activities such as conditioning exercises.

d. Fartlek Training

(l)Description and Function. Fartlek training is various running exercises conducted along an intense combat-like course. The running is conducted to overwork the lungs, allowing only partial recovery which is followed by another intense period of overload. This sequence is repeated for the duration of the workout. Fartlek training is a useful combat training method and general cardiovascular conditioner. The entire unit must be in uniformly good to excellent condition to qualify for this training.

(2) Progression. The following is an example of one fartlek session:

(a) Warm up by stretching 3-5 minutes.

(b) Warm up running easily 5-10 minutes.

(c) Run at a fast, steady pace for 3/4 to 1 3/4 miles (dependent on terrain).

(d) Walk/jog at a moderate pace for 5 minutes (recovery).

(e) Run easily sprinting 15 to 20 meters occasionally.

(f) Run full speed uphill for 175 to 200 meters.

(g) Warm down by running easily for 1/2 to 1 mile.

(h) Warm down by stretching 3 to 5 minutes.

(3)Leadership. To add variety, try incorporating combat movements, terrain association, a series of exercises or carrying table of equipment weapons (hit and roll). This will take some initiative in setting the course, but it allows for variety. In this manner, the fartlek course can be used as a training session or part of one. It works extremely well with highly motivated, competitive Marines.

3603.ORIENTEERING

Orienteering is land navigation over a prescribed course as a timed, competitive event. It is an excellent way to integrate land navigation training with physical conditioning and requires Marines to be skilled in both areas. Commanders adjust the training as needed to change the degree of difficulty or to emphasize either navigation or conditioning. Taken to the extreme, an orienteering exercise is conducted in full combat gear over rugged terrain against aggressors. Orienteering requires the following physical skills: endurance, speed, strength, and all-around conditioning.

Section VII. Circuit Training

3701. GENERAL

a. Description and Objective. A strength circuit consists of a series of stations where individuals in small groups exercise vigorously for a short period of time and then move (on signal) to the next station where a different form of exercise is conducted. This rotation continues until all groups move through all stations. Strength circuits contain no set or specific types of exercise stations within the circuit. The objective of circuit training is to develop strength. There are three general types of circuits.

(1) Fixed Circuit. This is a circuit in which apparatus of an immovable type (fixed into the ground) is used.

(2) Movable Circuit. This circuit consists of individual exercise apparatus which is portable and can be moved to and from the training area.

(3) Simplified Circuit. This circuit requires no equipment or apparatus.

b. Formation. The exercises are done at will, but rapid, steady, and continuous work is required of all. Each Marine's nervous and muscular system reacts differently to timed vigorous exercises. His performance should be measured on how many movements per exercise he can complete as an individual. For example, one Marine may be able to complete 5 movements, while another may be able to complete 20, and yet each is receiving the maximum benefit. All three circuits contained in this chapter are designed for platoon-sized groups. Expansion beyond this capacity requires a large amount of equipment, as each Marine in the fixed and movable types of circuits must have an item of equipment available for exercise at each station. A group larger than a platoon could be exercised through use of the simplified type of circuit; however, the group would be unwieldy and control could be a problem.

c. Place in the Program. All circuits illustrated can be completed in a 15-minute period. This feature allows the exercise of a platoonor smaller group on the circuit for a single 15-minute period, or the scheduling of the circuit as a 15-minute period within a longer period. A circuit can thus be utilized within a rotating activity system of scheduling. Choice of a circuit by the unit depends upon area, facilities, and other local factors; however, there is a circuit for every need.

3702. FIXED STRENGTH CIRCUIT

a. Description and Objective. The strength circuit is an arrangement of various types of exercise apparatus which are fixed in position. (See fig. 3-16.)Seven basic exercises are used and each exercise requires an apparatus. All apparatus of one type are positioned together to constitute a station. Each station will accommodate 10 Marines. The circuit is designed to be accomplished in 15 minutes when conducted on a time-rotation basis, as normally executed for unit training. For individual use, a Marine may complete the circuit by executing a specific number of repetitions for each exercise. In this case, the time required for completion of the circuit would vary slightly depending on the number of repetitions the Marine completed. The objective of this circuit is to provide a series of exercises which will improve and maintain the strength of the body's major muscle groups.

Figure 3-16 is available in the illustration section at the end of the book.

b. Starting Level. Marines must be thoroughly warmed up prior to participating in the circuit system. If Marines have not engaged in vigorous exercise immediately prior to starting the circuit, then an instructor should execute the following warm up exercises. These exercises should be conducted in the normal formation for set drills. Seven repetitions of each exercise will normally provide sufficient warm up. The exercises are-

(1) High jumper, Conditioning Drill 1.
(2) Bend and reach, Conditioning Drill 1.
(3) Squat bender, Conditioning Drill 1.

c. Progression. The instructor can adjust the circuit system through several methods to accommodate participants of varying physical ability. One method of adjusting an exercise is to change the method of executing the leg lift or to select a heavier weight for the barbell curls. When it becomes apparent during a unit program that the overall fitness of the group has Improved, then the exercise may be made more strenuous in two additional ways. First, the instructor can have one or more of the supplementary stations added. Second, the time spent exercising at each station can be increased in 5-second increments to a maximum of 60 seconds. Also, as Marines learn the circuit, the 45-second period for movement and instruction between stations can be eliminated, thus allowing only enough time to change stations.

d. Starting Positions. The Marine starts on any station, exercises steadily for a certain period (45 seconds initially), then moves on command to the next type of apparatus where he again exercises steadily for an equal period. The Marine continues until the required exercise is completed at each station. If it is desired to expand the number of stations in the circuit to accommodate more participants at one time, the instructor can provide four additional supplementary exercises, requiring no equipment. This will increase the amount of time required to complete the circuit.

e. Leadership. Close control of all Marines is necessary to ensure that a minimum amount of time is spent in moving them to their initial stations and in moving them between stations. One instructor can control the activity on the strength circuit. A stop watch or wrist watch with a second hand is required. When Marines arrive at the strength circuit, the instructor will form them for exercise and conduct the warm up drill. The group is then reassembled and formed into a number of files equal to the number of stations being used in the circuit. Each file is then directed to a station. As soon as all participants have reached an exercise position at a station, the instructor gives the command READY, GO. After 45 seconds of exercise, the instructor gives the command STOP, CHANGE OVER. The instructor allows Marines 45 seconds for moving to the next station and for preparing for the next exercise before the command EXERCISE is again given. In lieu of verbal commands, a whistle may be used to stop and start the exercises. For large groups, a megaphone or loudspeaker is useful.

f. Individual Conditioning Program. For the Marine working alone on the strength circuit, it would be impractical to time the exercise periods. For individual exercise, the Marine should select a number of repetitions of each exercise to complete, then rotate to the next stations after completing these repetitions. The number of repetitions selected should be at or near the maximum that the Marine is capable of doing without halting for rest.

g. Fixed Circuit Stations

(1) Primary Stations (See fig. 3-16.)

(a) Pullups/Chinups. A horizontal bar placed 8 feet above the ground is necessary for each Marine at this station. The Marine will also need a space on the bar that is 45 inches wide. On the command**EXERCISE**,the Marine grasps the bar with both palms facing either forward or to the rear, arms fully extended, and feet free of the ground and executes the exercise as described in MCO 6100.3_.The exercise is repeated as many times as possible until the command **STOP** is given. Then the Marine moves to the next exercise. If a Marine has done his maximum number of pullups/chinups prior to the command **STOP**, he will remain in the "dead" handing position until the command **STOP** and move to the next station.

(b) Twist Grip. The apparatus is a horizontal bar, free to turn, held between uprights placed 30 inches apart. The bar is 52 inches above the ground. A weight of 20 pounds is attached to the center of the bar by a light rope long enough to permit the weight to rest on the ground. The Marine stands at arm's length from the bar and grasps it with his hands on either side of the rope, palms down, thumbs under the bar. On the command **EXERCISE,** the hands are rotated so that the backs of the hands are rotated away from the body, thus winding the rope on the bar. The elbows are kept straight to ensure that the exercise is performed by the hand and forearm. When the weight is drawn up to the bar, the bar is then rotated in the opposite direction to lower the weight to the ground. This exercise is continued until the command **STOP** is given. The Marine then moves to the next station.

(c) Pulley Weights. The apparatus is a T frame with a system of pulleys that suspends a weight of about 90 pounds. The weight is attached to a light steel cable which has a drawbar attached to the other end. The Marine grasps the drawbar and sits down directly under the bar, legs extended to the front and arms extended overhead. The exercise is executed by pulling the drawbar down behind the head, then extending the arms slowly again until they are fully extended overhead. The exercise is repeated as many times as possible until the command STOP is given. The Marine then will move to the next exercise. Upon completion of the exercise, the weight is lowered slowly to the ground.

(d) Barbell Curls. A barbell is necessary for each Marine at this station. The barbell is constructed of 1 1/4-inch pipe 5 feet long, and two concrete-filled No. 10 cans. Each barbell should weigh about 40 pounds. Variance in the weight of the barbells, up to about55 pounds, will allow appropriate overload to be applied to Marines who are above average in strength or weight. The Marine grasps the bar with the palms forward and assumes a standing position with the barbell held in front of the hips, hands approximately shoulder's width apart. On the command **EXERCISE,** the Marine flexes the elbows and draws the barbell up until it touches the upper chest. The elbows remain at the sides. Breath is inhaled with the upward movement and exhaled as the barbell is lowered to the starting position. The exercise is repeated as many times as possible until the command **STOP** is given. The Marine moves to the next apparatus.

(e) Step-Up. The apparatus is a platform or ledge 18 inches high and of such size. At accommodate 10 Marines. The Marine faces the platform and on the command **EXERCISE,** steps up onto the platform, bringing his trailing foot up beside the leading foot. He then steps back down to the original position, stepping down first with the same foot he initially used in stepping up. After 10 repetitions of the exercise, he changes the order of moving the feet to use the opposite leg for stepping up. Repeat this exercise until the command **STOP** is given. The Marine then moves to the next station.

(f) Leg Lift. The apparatus is a horizontal bar constructed as described in (a). To prevent the body from swaying, a horizontal back support is added 40 inches below the horizontal bar. The arms are kept fully extended. On the command **EXERCISE,** the Marine jumps up, grasps the bar with the palms forward and the back support behind him. The exercise is executed by raising the legs to a horizontal position then slowly lowering them to the vertical position. The Marine does not flex his knees. He does not swing the legs to the rear of a vertical position to gain momentum for raising them in the next repetition of the exercise. The movement is repeated until the command **STOP** is given. The Marine then moves to the next exercise. If unable to raise his legs to a horizontal position without flexing his knees, the Marine flexes his knees and draws them up to the chest, then lowers his legs to the vertical position.

(g) Rope Climb. The rope climb is 20 to 30 feet high with five ropes suspended from a horizontal bar which forms the uppermost part of the framework. To prevent the horizontal bar from sagging and to provide safety, only five ropes are attached to it. There are two frameworks per station. The ropes are 6 feet apart. Any method may be used to climb the rope, and the Marines climb as high as possible. Marines who are proficient should climb therope several times during the time allotted. Inexperienced Marines should be cautioned to take care during descent to avoid rope burns on their hands.

(2) Supplementary Stations. The following exercises are designed to expand the basic circuit by being inserted in specific places within the system. For each supplementary station used, there should be adequate room for 10 Marines to exercise.

(a) Bent Knee Situp or Bottoms-Up. These calisthenics are designed to strengthen the abdominal muscles. These exercises will be inserted between the pullup/chinup and twist grip stations. The primary stomach exercise is the situp. In case of inclement weather or other conditions that make ground contact undesirable, the bottoms up exercise is used. In situps on the command **EXERCISE,** the Marine lies on his back (supine position) with knees flexed and both feet flat on the ground and executes the exercise (minus an assistant) as described in MCO 6100.3_.The Marine then returns to the starting position, repeating the exercise until the command **STOP** is given. The Marine then moves to the next station. In bottoms up on the commandEXERCISE,the Marine assumes the front leaning rest and executes the bottoms-up exercise as described in Conditioning Drill 3. He continues this exercise at a moderate cadence until the command **STOP** is given. The Marine then moves to the next station.

(b) Pushup. This exercise is designed to strengthen the arm- and shoulder-girdle muscles. It should be included between the twist grip and pulley weight stations. Upon the command **EXERCISE,** the Marine executes the pushup as described in Conditioning Drill 1. He continues this exercise at a moderate cadence until the command **STOP** is given. The Marine then moves to the next station.

(c) Knee Bender. This exercise is designed to build leg muscles and is included between the pulley weight and barbell curl stations. On the command **EXERCISE,** the Marine executes the knee bender as described in Conditioning Drill 3.He continues this exercise at a moderate cadence until the command **STOP** is given. The Marine then moves to the next station.

(d) Trunk Twister. This exercise strengthens the major muscles of the trunk and is included between the step-up and pullup/chinup stations. On the command EXERCISE, the Marine executes the trunk twister as described in Conditioning Drill 1. He continues this exercise at a moderate cadence until the command **STOP** is given. The Marine then moves to the next station.

3703. MOVEABLE STRENGTH CIRCUIT

a. Description and Objective. The exercises in this circuit are progressive and the course is planned to gain and hold the interest of the participating groups. The circuit consists of a series of stations, with each station designed to develop a particular group of muscles. Along with muscular development, correct posture and deep rhythmic breathing should be stressed on this circuit at all times.(See figs. 3-17 and 3-18.)

Figure 3-17 is available in the illustration section at the end of the book.

b. Equipment. The equipment is set up in files. Six files of 8 stations will accommodate a platoon

c. Formation. The platoon marches to the area where the equipment is positioned and forms a file within each lane of stations, covering down on a piece of equipment. Movements are made on the double, the important factor being that no time is wasted in getting to work.

d. Progression. Initially 40 to 45 seconds per station is adequate. As individuals become stronger the time should be increased in 5-second increments until a minute to a minute and a half is reached.

e. Leadership

(1) The leader stands in front of the barbell station and controls the rotation from this position. The leader supervises the entire group, with the assistance of several instructors who move about in the platoon correcting and encouraging the Marines.

(2) The leader starts each group but does not count cadence nor lead them through the exercises. Each individual exercises rapidly but individually.

(3) As each Marine finishes his repetitions with the barbell, he places the barbells on the ground. The leader calls, READY, followed by the command, FALL OUT ONE. All will double-time to the station directly in front of them, while the individuals on the barbell stations will do an about face to the rear station in their lane.

f. Movable Circuit Stations. The Marine can obtain best results on the movable circuit if the exercises on the various stations are given in the following manner:

(1) Station 1—The Barbell. (See fig. 3-17, 1.)The exercises at this station stress proper posture; deep, rhythmic breathing; and development of the muscles of the arms, shoulders, and upper body. An instructor teaches the proper methods of lifting before the exercises begin. The methods are to lift with the legs, to keep the back straight, and to merely grip with the hands. Two recommended exercises are given below. The instructor will use only one exercise per period .He may use either exercise. At successive periods, the other exercise is used.

(a) Exercise 1--Two-Hands Military Press. (See fig. 3-18, 1.)Grasp the barbell with both hands, knuckles up at shoulder width, and lift to the chest. Steadily press to arm's length overhead; lower to the chest resisting weight all the way. Inhale as the weight is pressed up and exhale as the weight is brought down.

(b) Exercise 2--Two-Hands Regular Curl. (See fig. 3-18, 1.) Lift the weight to the waist, with the palms of the hands out, heels together, stomach in, chest lifted and arched, shoulders back, elbows in close to the sides; inhale deeply and curl the weight to the shoulders, using the arms only, at the same time keeping the elbows close to the sides; exhale Emphasize posture and the use of the arms only. This is a very valuable exercise for the development of the biceps and the grip and should be repeated from 8 to 16 times, depending on the ability of the Marine.

(2) Station 2--The Jump Rope. (See fig. 3-18, 2.) This exercise develops strength and agility in the legs and stamina of the whole body. It makes the Marine agile on his feet and increases his footwork efficiency and timing. Each Marine should progress until able to jump rope at least 3 minutes at top speed.

(3) Station 3--The Twist Grip. (See fig. 3-18, 3.) The twist grip is an excellent exercise for the hands and forearms, and adds greatly to the Marine's ability in hand-to-hand combat. The handle is gripped and twisted, winding the rope until the weighted can is level with the height of the hands, which are held horizontal. The weight is lowered in the same manner; the Marine resists the weight all the way, occasionally stopping the twisting motion and alternately removing first one hand, then the other, from the handle. A variation of the above exercise is to wind the handle with the palms up and the arms bent and with the elbows held close in to the sides. Each Marine should maintain a good posture and keep the stomach muscles taut throughout this exercise.

(4) Station 4--The Incline Plane. (See fig. 3-18, 4,)The use of the incline plane is a very strenuous exercise and well designed for the development of the abdomen. Although 6 to 10 repetitions are sufficient for the beginners, more can be added as ability increases. Marines with hernias or recent operations will be excused from participation at this station.

(5) Station 5--The War Club. (See fig. 3-18, 5.) The war club is a simple and effective means of exercising the principal muscle groups of the body, especially those of the trunk, back, and shoulders. To gain the maximum benefit from this exercise, the Marine must keep both feet flat on the ground at all times. Throughout the exercise period, the weight is swung from arm's length as follows:

(a) As in chopping wood, first on one side, then on the other.

(b) As a batter warming up with a number of bats.

(c) In large circles, first with one hand and then with the other.

(6) Station 6--The Bicycle Ride. (See fig. 3-18, 6.) The bicycle ride is well suited to exercising many of the muscle groups of the body, particularly those of the abdomen. Vary the speed of the exercise, but keep everyone "riding" the entire period. A variation exercise may be performed by placing the legs together, raising them slowly to a height about 2 feet from the ground, and then lowering them slowly to the ground.

(7) Station 7--The Step-up. (See fig. 3-18, 7.) The step-up exercises the legs. The step-up is performed by initially stepping up with the left foot, followed by the right, then stepping down with the left foot followed by the right. Continue for 20 seconds, then change to the right foot as the lead foot for 20 seconds.

(8) Station 8--The Isometric Pull. (See fig. 3-18, 8.) Two trainees work at this station with a cable pull and perform the following exercises:

(a) Initially start with one Marine in the supine position and one Marine sitting. The sitting Marine lowers the upper body to the ground and pulls the partner up to the sitting position. The partner then performs this same action and this is continued for 20 seconds at a rapid rate.

(b) During the last 20 seconds, the same action takes place but, in this case, the Marine in the supine position resists the pull of the partner for approximately 5 seconds before allowing to be pulled up into the sitting position.

3704. CIRCUIT-INTERVAL TABLE

a. Description and Objective. The circuit-interval table is designed to develop strength and endurance within a short period of time, with no equipment requirement, through a rapid and vigorous routine of exercise. Fifteen minutes is an adequate period to execute all exercises and to secure a vigorous workout with the circuit-interval principle.

Figure 3-18 is available in the illustration section at the end of the book.

b. Formation and Starting Positions

(1) A leader forms platoons or smaller groups in an oval or circular formation with 3- to 5-yard intervals between Marines. The Marines face to the right and move forward at quick time and then double time. (See fig. 3-19, A.)After running several platoon circle laps, the leader calls out the name of an arm and shoulder exercise from the list below, orders quick time and commands, for example, PUSHUPS. On this command, all Marines immediately hit the ground and individually and rapidly begin doing pushups. No cadence is counted. (See fig. 3-19, B.) After 30 seconds of exercise, the leader commands. **QN YOUR FEET, FORWARD, MARCH** The platoon resumes the quick time cadence and the leader, when ready, gives the necessary commands for double time. The double time is continued for one or more laps and the leader calls out the name of the next exercise and the process is repeated. This continues, with running between each exercise, until every body part has been exercised.

(2) The instructor controls the running and quick time to observe the effects of the exercise upon the Marines. Cadence, step, and precision are not important to the objective and the instructor should not use them. What is important is speed and the instructor should stress this. After the exercise period is started, the Marines do not stop. This circuit method emphasizes stress and recovery, the recovery occurring during the quick time periods.

c. Activities. The leader can use the following exercises. These exercises can be repeated if necessary, during a second round. The leader can use other calisthenic exercises in the circuit-interval table.

(1) Arms and shoulders-push-ups.

(2) Stomach situps.

(3) Back-squat thrusts.

(4) Legs-bicycle (on back).

d. Progression. The progress is controlled by the leader, who must pace the running, quick-time movement, and exercise in such a way that Marines will receive a vigorous workout yet be able to participate throughout the 15-minute period. Marines who are in the initial stages of physical condition will not be able to double time or exercise as long as those who are better conditioned. The idea is to set a pace which can be increased during each workout, thus progressing gradually to a higher level of physical fitness.

e. Leadership. The platoon leader, platoon sergeant, or section leader can lead the group. The leader must execute the exercise with the unit to feel the effects and thereby adjust the pace.

f. Place in the Program. This activity may be scheduled whenever a short period of time is available. The only requirement is that enough space, indoors or out, be available to form the circle.

Figure 3-19 is available in the illustration section at the end of the book.

Section VIII. Basic Physical Skills and Obstacle Courses

3801. GENERAL

The purpose of this section is to list the basic military skills and the methods for their development. Many of these skills are best developed by obstacle courses but other drills are also discussed. The section explains types of obstacle courses, details of construction, and methods of negotiating the various obstacles.

3802. BASIC PHYSICAL SKILLS

a. Objective. The objective is to develop proficiency in the various military physical skills which are essential to personal safety and effective combat operations. In travel by foot over rugged terrain and in the execution of combat duties, Marines must be trained to perform basic skills, such as running, jumping, climbing, and carrying. During training, Marines will develop agility and coordination in these skills .Fast and skillful execution of these skills may mean the difference between success and failure on the battlefield.

b. Place in the Program. These skills are practiced throughout the entire program in many different activities. Many of these skills are best practiced on obstacle courses.(See par. 3803.)

c. Description. These basic skills are the minimum skills required by all Marines. The basic skills are as follows:

(1) Running. Running is used to strengthen the legs and develop cardiovascular endurance. Marines should be exposed to running in various situations: on roads, over rough ground, up and down hills, cross-country, and running over low obstacles.

(2) Jumping. In broad jumping, the take-off foot is planted firmly and the spring comes from the extension of this leg as the other leg reaches for the far side of a ditch or similar obstacle. (See fig. 3-20.) The arms are forcibly raised forward and upward to assist in propelling the body up and forward. Landing may be on one or both feet depending upon the length of the jump. In vertical jumping downward from a height, the jumper should aim his feet at the desired landing spot and he should also jump with the knees slightly bent and feet together, with the trunk inclined slightly forward. As the feet touch the ground, the shock is absorbed by bending the knees into a full squatting position. If the height is too great or the ground too hard to absorb the shock, then the jumper should forward roll or side roll thus eliminating some of the momentum.

Figure 3-20 is available in the illustration section at the end of the book.

(3) Dodging In combat situations, it is often times necessary to change directions quickly. To execute this movement while running, a lead foot is firmly planted, left foot if the direction is to the right and right foot if the direction is to the left. The opposite foot is moved toward the new direction. The knees are slightly flexed during the movement and the center of gravity is low and balanced. (See fig. 3-21.) At the time of the change of direction, the head and trunk are turned quickly in the new direction.

Figure 3-21 is available in the illustration section at the end of the book.

(4) Climbing and Surmounting. All Marines should know how to climb and surmount various types of obstacles. (See fig. 3-22.)

Figure 3-22 is available in the illustration section at the end of the book.

(a) Vertical Climbing, as in Climbing a Rope or Pole. The hands grasp the rope or pole overhead with the palms toward the face. Gripping the object, the body is pulled upward with the arms and shoulders, assisted by the feet which grip the object and assist by pushing downward. If shoulder girdle strength and body coordination are not adequate to permit alternating the hands, the arms act together in pulling upward.

(b) Climbing as in Surmounting a Wall. Ingoing over a wall, the body should be kept as close to the top as possible, since in combat operations it is important to offer as small a target as possible to the enemy. If an individual climbs a wall while carrying a rifle, both hands should be freed by slinging the rifle over the back. There are two methods commonly used for surmounting a wall of moderate height, but only one for dropping from it.The methods are as follows:

1 Run, Jump, and Vault. Approach the wall at a run, jump forward and upward at the wall and place one foot against it as high up as possible. Use the foot in contact with the wall to help push the body upward while grasping the top of the wall with the hands. Pull the body up with the arms, assisted by pressure of the foot against the wall and swing the legs over, propelling the body weight over the wall.

2 Hook and Swing. Approach the wall at a run and jump forward and upward. Hook one elbow over the wall, locking the arm in place by pulling up until the top of the wall is underneath the armpit. Grasp the top of the wall with the other hand. Draw the leg which is closer to the wall up as far toward the top as possible. Then swing the other leg over the top of the wall.The body is then carried over with a rolling motion. A variation of this leg action can be used by Marines who are unable to draw up the leg as described. While hanging with both legs fully extended, start a swinging motion with the legs together. When the legs have enough momentum, swing the outside leg over the top of the wall with a vigorous kick, then follow with the body.

3 **Dropping.** All drops from a height are executed in the same manner regardless of the method used to gain the top. One hand is placed against the far side of the wall while the other hand grasps the top. From this position, the body is rolled over the wall and vaulted away from it with the legs swinging clear. As the body passes over the wall and drops, it should at all times face the wall. This will keep the rifle and other equipment clear. Break the fall by retaining a grasp on the top of the wall as long as possible.

(c) Climbing Ladders and Cargo Nets. Rope ladders, stationary vertical ladders, and cargo nets employ the same general technique. The important element is to grasp the side supports firmly in the hands about shoulder height and place the feet on a rung which would cause the body to be fully extended. In movement upward, one hand is moved upward and a new grasp is secured and, at the same time, the opposite leg moves up a rung. As the knee straightens, the body is elevated. This process is repeated using the opposite arm and leg. Alternation continues in this manner until the climber reaches the objective.

(5) Traversing Horizontal Objects. The traversing of horizontal objects puts heavy stress on the arms and shoulder girdle area as the feet are usually suspended in the air with all of the body weight on the arms and shoulders.(See fig. 3-23.)

(a) Traversing Horizontal Ropes or Pipes. The hands grasp the horizontal support overhead with the palms facing. To propel the body forward, one hand is released and moved forward to secure a new grasp. At the same time, the opposite side of the body is swung forward (some people are able to "walk" in the air, keeping the body to the front and moving the legs in time with the arms as in walking on the ground). The other hand is then released and moved forward; this alternation is continued until the objective is reached.

Figure 3-23 is available in the illustration section at the end of the book.

(b) Traversing Horizontal Ladders. In this situation, the movement is the same as used in traversing a rope or pipe. The hands, however, are placed on the rung with the palms away from the face. Other than this difference the technique is the same.

(6) Crawling. Crawling in combat situations is an often used skill. Crawling may be high or low.(See fig. 3-24.)

Figure 3-24 is available in the illustration section at the end of the book.

(a) High Crawl.

In the high crawl, the Marine moves on hands and knees, moving one hand and the opposite knee and then continuing to move the hands in alternation with the opposite knee following the companion hand.

(b) Low Crawl. The Marine is in the prone position usually with the forearms and palms of the hand on the ground. He propels forward by bending the knee of one leg and pushing with the inside edge of the shoe. At the same time, the opposite arm moves forward and pulls to the rear. The body remains low and movement is continued by bending the opposite knee and pushing, and at the same time sliding the opposite arm forward and pulling. Alternation of hands and legs continues until the objective is reached.

(7) Throwing. Throwing may be executed from the kneeling or standing positions. The object to be thrown is held in the hand, and the throwing arm is bent at the elbow; the hand is then moved to the rear until the hand is behind the ear. The body is turned so that the lead foot and balance arm on the side toward the target are pointing at the target. (See fig. 3-25.) The balance arm is used to sight over and align the throwing hand and the target. When properly aligned, the elbow is moved rapidly forward until it is at a point just in front of the body where the arm is straightened and the wrist snapped. This whip motion propels the object to the target. Underhand throws secure momentum by the thrower bending his knees and swinging the throwing arm to the rear. As the knees are straightened, the arm is forcefully swung forward from the shoulder and the object released.

Figure 3-25 is available in the illustration section at the end of the book.

(8) Vaulting. Vaulting is employed to overcome low barriers or fences. (See fig. 3-26.)The object to be surmounted is approached at an angle. The hand on the side next to the obstacle is placed on the top of the obstacle and, with a straight arm, the body weight is pushed upward. At the same time, the leg on the side next to the obstacle is thrown upward and over the top followed by the other leg (side approach). In landing, the weight comes down on the leading leg first followed by regaining the balance on both legs. The free arm serves as a balance. A direct (front) approach can be used at which time both legs go over the object together.

(9) Man Carrying. There are three basic individual means of carrying personnel in combat situations and one of these methods may be used in carrying objects.

Figure 3-26 is available in the illustration section at the end of the book.

(a) Fireman's Carry."A" stands sideways in front of "B", "A" bends his knees and leans forward, placing one arm through "B's" crotch, grasps the wrist of "B's" arm, which is hanging over the shoulder, and then "A" runs forward.

(b) Saddle-Back Carry. With his back toward "B", "A" stands in front of "B". "B" mounts "A's" hips and clasps his arms in front of "A's" chest. "A" grasps "B's" thighs.

(c) Single-Shoulder Carry. "A" stands facing "B"."A" assumes a semi-squatting position. "B" leans forward until "B" lies across "A's" right shoulder. "A" clasps his arms around "B's" legs and straightens up, lifting "B" from the ground. "A" then runs forward. This method may also be used to carry heavy objects.

(10) Balancing. Balancing the body while walking or running on a narrow object when crossing obstacles is a skill which requires practice and confidence. Balance is required in negotiating a log placed across a stream, in crossing a narrow beam or rail, and in similar situations.(See fig. 3-27.) To perform this skill, place the feet on the object to be crossed, hold the arms to the sides at shoulder level, and fix the eyes on the object approximately 5 yards in front of the feet. Generally, it is not a good practice to look down at the feet. Walk the beam by placing first one foot and then the other in the center of the beam, thereby moving forward, using the arms to aid in maintaining balance.

Figure 3-27 is available in the illustration section at the end of the book.

(11) Falling. Injury can be avoided if Marines are taught to fall properly. They should know how to use the body momentum to their advantage during a fall rather than to try resisting that force. (See fig. 3-28.) If enough force is present, such as occurs during a fall while running or in jumping downward from a height, individuals can extend their hands to catch the weight. At the same time, duck the head and roll forward onto the feet. The key to falling without injury from the standing position is relaxation and rolling on the outside of the leg, hip, and buttocks to take the brunt of the fall.

Figure 3-28 is available in the illustration section at the end of the book.

3803. OBSTACLE COURSES

a. General. Obstacle courses are a valuable part of physical readiness training. The challenge presented by the obstacles assists in developing and testing the basic physical skills. In many combat situations, success will depend upon the Marine's ability to perform one or more of these skills, often while carrying field equipment and when fatigued. In this section, two types of obstacle courses will be discussed-Conditioning Obstacle Course and Confidence Obstacle Course.

b. Course Safety Precautions. Commanders and course instructors should take certain precautions to prevent injury on obstacle courses. A few of the precautions are:
(1) Inspect the course for faulty construction of obstacles, protruding nails, rotten logs, condition of the landing pits, and other hazards to safety.
(2) Conduct warm up exercises before the unit runs the course.
(3) Explain and demonstrate the correct techniques for negotiating all the obstacles before allowing the Marines to try them.
(4) Give Marines at least two weeks of conditioning exercises before scheduling the obstacle and/or confidence courses.
(5) Ensure that negotiation of the higher and more difficult obstacles are under the supervision of an instructor.
(6) Do not permit individuals who have neither practiced the basic skills nor run the conditioning obstacle course to participate in the confidence obstacle course.
(7) Weather conditions may cause footing or handhold surfaces to be slippery. If such is the case, postpone training on the course.

3804.CONDITIONING OBSTACLE COURSES

a. Description and Objective. The Conditioning Obstacle Course is commonly known as the Obstacle Course. This course consists of fairly low obstacles which are designed to be negotiated quickly. The obstacles serve to test various basic skills, and running the course is a test of the Marine's physical condition. After receiving instruction and an opportunity to practice the skills, Marines run the course against time.

b. Area and Equipment
(1) Complete standardization of obstacle courses should not be attempted since topographical conditions always vary. Commanders should use ingenuity in constructing a course, making good use of streams, hills, trees, rocks, and other natural obstacles. Since the course is eventually run at high speed, it should not be dangerous.

(2) The course should be wide enough for six or eight men to run simultaneously, encouraging competition. The lanes for the first several obstacles should be wider and the obstacles themselves easier than those that follow. This avoids congestion until the contestants scatter out over the course. The last two or three obstacles should not be too difficult and should not involve high climbing. This prevents injuries and falls resulting from fatigue.

(3) The total distance of the course should range from 300 to 450 yards and include from 15 to 25 obstacles. Normally the obstacles should be 20 to 30 yards apart and arranged so that those which exercise the same groups of muscles are separated.

(4) The obstacles should be substantially built. Peeled logs, 6 to 8 inches in diameter, are ideal for many of the obstacles. Sharp points and corners should be eliminated. Landing pits for jumps or vaults should be filled with sand or sawdust to prevent injuries.

(5) The course should be constructed and marked so that it is not possible to sidestep or detour obstacles. However, it is desirable to provide alternate obstacles of varying degrees of difficulty.

(6) The course should be in the shape of a horseshoe or figure eight so that the finish is close to the start and signs should be placed to indicate the course route.

c. Leadership. Before Marines run an obstacle course, they should be instructed in the proper technique of negotiating each obstacle. In each case this technique should be explained and demonstrated in detail, with emphasis on avoiding injury. Every individual should be given an opportunity to practice on each obstacle until he becomes reasonably proficient at negotiating it. Before the course is run against time, it is advisable for individuals to make several runs at a slower pace. During such practice or trial runs, the instructor should observe the performances and make appropriate corrections. Marines should never be permitted to run the course for time until they have practiced on all obstacles. The best method of timing the runners is to have the timer stand at the finish and call out the minutes and seconds as each individual finishes. If several watches are available, each wave may be timed separately. If only one watch is available, the different waves should be started at regular intervals, such as every 30 seconds. If an individual fails to negotiate an obstacle, a previously determined penalty should be exacted.

d. Types of Obstacles

(1) Jumping-Type Obstacles. These obstacles may be ditches which are cleared with one leap, trenches which the individuals can jump into, heights which require jumping downward, or hurdles which an individual can jump over. (See fig. 3-29.)

Figure 3-29 is available in the illustration section at the end of the book.

(2) Dodging-Type Obstacles. Obstacles of this type are usually mazes consisting of posts set in the ground at irregular intervals. (See fig. 3-30.)The intervals between posts should be rather narrow so that the Marines must pick their way carefully through and around them. Lane guides may be constructed to guide the Marines to dodge and change direction. Obstacles may be put into a maze pattern to cause the Marines to change direction.

Figure 3-30 is available in the illustration section at the end of the book.

(3) Vertical Climbing and Surmounting Type Obstacles. These obstacles may be climbing ropes, either plain or knotted and 1-5 inches in diameter; cargo nets or walls 7 or 8 feet high; or vertical poles 6 to 8 inches in diameter and 38 feet high.(Seefig. 3-31.)

Figure 3-31 is available in the illustration section at the end of the book.

(4) Horizontal Traversing-Type Obstacles. Horizontal obstacles may be pipes, beams, ladders, or ropes. (See fig. 3-32.)

(5) Crawling Type Obstacles. Obstacles which require crawling may be constructed of large pipe sections, low rails, and wire. (See fig. 3-33.)

Figure 3-32 is available in the illustration section at the end of the book.

(6) Vaulting-Type Obstacles. Obstacles of 3 to 3.5 feet in height such as low walls or fences may be used as a vaulting obstacle.(See fig. 3-34.)

(7)Balancing-Type Obstacles. Beams, logs, and planks may be used as balancing-type obstacles.(See fig. 3-35.)These items may be used to span water obstacles and dry ditches, or raised off the ground somewhat to simulate natural depressions.

Figure 3-33 is available in the illustration section at the end of the book.

3805.CONFIDENCE OBSTACLE COURSES

a. Description and Objective. The Confidence Obstacle Course is commonly known as the Confidence Course. This course is composed of higher and more difficult obstacles than those used in the conditioning course. The confidence obstacle course is designed to give Marines confidence in their mental and physical capacities and to cultivate their spirit of daring. They are encouraged but not compelled to negotiate this course and the course is not run against time. The negotiation of a confidence course, however, is strenuous enough to be an excellent physical conditioner. Marines should NEVER attempt to take the obstacles at high speed and should not compete for speed. The obstacles vary from fairly easy to extremely difficult. Some are of considerable height to accustom Marines to climbing such heights without fear. Considerable emphasis is placed on obstacles that train and test an individual's balance.

Figure 3-34 is available in the illustration section at the end of the book.

b. Area and Equipment

(1) The confidence course accommodates four platoons, one platoon at each group of six obstacles. The course should be made up of about 24 obstacles, numbered and marked as follows:! to 6, white numbers on red background; 7 to 12, black numbers on a white background;

13 to 18, white numbers on a blue background; and 19 to 24, white numbers on a black background.

(2)A few simple pieces of equipment should be provided for individuals who do not have the strength, courage, or ability to negotiate the obstacles.

Figure 3-35 is available in the illustration section at the end of the book.

c. Formation. The obstacles should be divided into groups of six, and each group is designated by a different color. Each platoon starts at a different color. Individuals are separated into groups of 8 to 12 at each obstacle. At the starting signal from the company commander, they proceed numerically through. Anyone may skip an obstacle who is afraid to try. Individuals proceed from obstacle to obstacle until time is called, then assemble as ordered.

d. Leadership. If the Marines are new to the confidence course, an instructor will demonstrate or will give a brief orientation at each obstacle, including an explanation and demonstration of a method of negotiating it. Marines are encouraged to try the various obstacles, but they are not compelled to do so. No compulsion is to be used. The manner of negotiating any obstacle is left to the discretion of the Marine. However, the instructor assists anyone who experiences difficulty. Instructors must supervise closely at all times to prevent injuries, as some of the obstacles are quite high. Also, some of the obstacles should not be used when slippery or wet. The example of instructors and especially selected demonstrators will serve to inspire the individuals to greater effort.

e. Negotiating the Obstacles. Although personnel need not conform to any one method of negotiating the obstacles, there should be some uniformity in the approach to them. A general method of negotiating the obstacles is indicated below.

(1) Red Group This group contains the first six obstacles. (See fig. 3-36.)

(a) The Belly Buster Individuals may vault, jump, or climb over. Warn them that the log is not stationary.

(b) Reverse Climb. Climb the reverse incline and go down the other side to the ground.

(c) The Weaver. Move from one end of the obstacle to the other by weaving the body under one bar and over the next.

(d) Hip-Hip. Step over each bar, either alternating legs or using same lead leg each time.

(e) Balancing Logs. Step up on log, retaining the balance, walk or run along log.

(f) Island Hopper. Jump from one log to another until the obstacle is negotiated.

(2)White Group. This group is composed of the second six obstacles. (See fig. 3-37.)

(a) Tough Nut. Step over each X in the lane.

(b) Slide for Life. Climb the tower, grasp the rope firmly, and swing the legs upward. Hold the rope with the legs to distribute the weight between them and the arms. Braking the slide with the feet and legs, proceed down the rope. Warn Marines that there is danger of getting rope burns on their hands. When the rope is slippery or wet, this can be a dangerous obstacle.

(c) **Low Belly Over.** .Mount the low log and jump onto the high log, both arms grasping over the top of the log, the stomach area in contact with it. Swing the legs over the log and lower the body to the ground.

(d) **Belly Crawl** Move forward under the wire, belly down, to the end of the obstacle.

(e) **The Dirty Name** Mount the low log and jump to or reach the higher logs in succession, then jump or drop to the ground. Warn the Marines about the height of the final log.

(f) **The Tarzan** Mount the lower log and walk the length of it and each successive, higher log until reaching the horizontal ladder. Grasp two rungs of the ladder and swing the body into the air. Negotiate the length of the ladder by releasing one hand at a time and swing forward, grasping a more distant rung.

(3) **Blue Group.** This group is formed by the third group of six obstacles. (See fig. 3-38.)

(a) **High Step over**. Step over each log, alternating the lead foot or using the same lead foot.

(b) **Swinger.** Climb onto the swinging log and over to the ground on the opposite side.

(c) **Low Wire.** Move under the wire on the back, using the hands to raise the wire to clear the body.

(d) **Swing,** Stop and Jump. Gain momentum with a short run, grasp the rope, and swing the body forward to the top of the wall. Release the rope while standing on the wall and jump to the ground.

(e) **Six Vaults.** Vault over the logs, using one or both hands.

(f) **Easy Balancer.** Walk up one inclined log and down the one on the other side to the ground.

(4) **Black Group.** The last group is formed by the final six obstacles. (See fig. 3- 39.)

(a) **Incline Wall.** Approach the underside of the wall, jump up and grasp the top and pull the body up and over. Slide or jump down the incline to the ground.

(b) **Skyscraper**. Jump or climb to the first floor, climb up the corner posts or assist each other to any desired floor. Descend to the ground in any desired manner.

(c) **Jump and Land**. Climb up the ladder to the platform and jump to the ground.

(d) **Confidence Climb**. Climb the inclined ladder to the vertical ladder. Go to the top of the vertical ladder, then down the other side to the ground.

(e) **Belly Robber.** Step on the lower log and assume the prone position on the horizontal logs. Crawl over the logs to the opposite end of the obstacle.

(f) **The Tough One** Climb the rope or pole on the higher end of the obstacle, then go down the ladder and across the log platform. Climb over or between the logs at the end and go down the rope or pole to the ground. Vault over the final log.

Figure 3-36 is available in the illustration section at the end of the book.

Figure 3-37 is available in the illustration section at the end of the book.

Figure 3-38 is available in the illustration section at the end of the book.

Figure 3-39 is available in the illustration section at the end of the book.

Section IX Individual Exercise Programs

3901.GENERAL

Marines at times will **be** stationed on independent duty and consequently will be responsible for their own physical fitness program. This section will assist them in understanding the need for exercise and will aid in the planning and execution of an individual exercise program. Exercise activities included are the Bench Conditioner, 6-12 Plan, Weight Training, and Isometric Contraction. Keeping physically fit is a problem that faces every Marine. Even though we are frequently engaged in training that requires some physical effort, in many cases, it is not enough to prepare us to meet the intense physical demands of combat. Attaining a satisfactory level of physical readiness is not an insurmountable objective. Available time appears to be the most difficult obstacle to the development of physical readiness. In most cases, regular physical training programs are centralized, requiring the individual to temporarily leave the work area. The problems involved in setting an hour aside two or three times each are numerous. However, most of us can devote a few minutes each day to physical fitness with little, if any, impact on our daily work schedule, especially if it does not require us to leave our work area.

a. Type of Program. There are many good physical fitness programs available to the individual or group. Regardless of the type or duration, to be effective, the program must contain exercises that are strenuous and are challenging to the individual. Space will not permit the inclusion of all available means of individual exercise. The programs selected for this chapter have met the requirement of minimum space and minimum time.

b. Need to Augment Program. These programs are quite strenuous and will develop a satisfactory level of physical readiness. However, if the Marine desires additional development of endurance, it is recommended that he supplement these programs with a 15- to 30-minute period of wind sprints and double timing on an alternating daily basis.

c. Progressive Training. If Marines are performing duties which require little or no physical activity, they must plan a physical conditioning program that assures a moderate beginning, moderate but steady progression, and sufficient warmup before starting the vigorous exercise. To avoid organic of bodily harm, a Marine activity without adequate warmup. He should conduct conditioning programs on a daily basis over an extended period of time, never on an unduly accelerated or crash basis.

3902.THE BENCH CONDITIONER

a. Description and Objective. The bench conditioning program uses a modified bench to employ both isotonic (moving) and isometric (stationary) exercises are as the nucleus of the program. The exercises are designed to develop strength and endurance in all the major muscle groups of the body. The principles of progression, overload, and balance are employed when the exercises are performed properly.

b. Area and Equipment. The conditioning apparatus can be constructed in any unit motor pool with welding equipment found in most salvage yards.(See figs. 3-40, 3-41.)Additionally, there are available any number of commercially produced apparatus available. The important thing is that in utilizing this equipment the following exercise routine be adhered to.

c. Starting Level and Progression.. The program consists of two tables, each with 10 exercises. The Marine controls progression by required repetitions or, in some cases, by application of maximum effort. Each table can be completed within 15 minutes

d. Starting Position. To start the program, the Marine begins with the Table 1 and executes each exercise for the required number of repetitions as indicated. The Marine controls the starting level and progression. When the Marine executes the maximum repetitions for Table I within a 15-minute exercise period, he progresses to Table II. To maintain this level of development, the Marine should also execute the maximum repetitions for Table II within a 15-minute exercise period. The Marine should keep substitution of exercises to a minimum. However, it he completes a full 15 minutes of strenuous exercise and exercises all muscles, then there should be no appreciable difference in the overall development.

e. Bench Conditioning, Table I

(1) Exercise 1: Side-Straddle Hop. This is a two-count warmup exercise done at moderate cadence. The starting position is the position of attention. On count ONE jump slightly into the air, swinging the arms out to the sides and up to a vertical position, hands touching.(See fig. 3-40, A.)At the same time, spread the feet wider than shoulder-width apart. On count TWO, using a slight flexing of the knees and ankles, jump slightly into the air and return to the starting position by swinging the arms back down to the sides. Twenty repetitions of this exercise is the standard dosage throughout the program.

(2) Exercise 2: Hand Walk. Remove the lower horizontal bar. Adjust the upper horizontal bar so that it is high enough to permit a "dead hanging" position, with the feet off the ground.(See fig. 3-40, B.j From the "dead hanging" position, release one hand and drop the arm to the side of the body. Then raise that arm and regrasp the horizontal bar. Release the bar with the other hand and drop that arm to the side. Repeat this as many times as possible.

(3) Exercise 3: Situps. Lie down with the fingers interlocked and placed behind the head. Hook the toes under the foot braces. Raise the trunk and upper body to an upright sitting position, twisting it to the left and then forward and downward until the right elbow touches the left knee.(See fig. 3-40, C.)Lower the body to the starting position .Sit up again but twist the body to the opposite direction as before, touching the left elbow to the right knee. Again lower the body to the starting position. The starting dosage is 20 situps. The Marine should continue the progression until he has attained 40 situps.

Figure 3-40 is available in the illustration section at the end of the book.

(4) Exercise 4: Double Step-upStarting at one end of the bench, step up onto the bench, and walk across it. Step down from the other end; turn around and repeat the process to return to the starting point.(See fig. 3-40, D.)Each return to the starting point constitutes a repetition. The starting dosage is 20 repetitions. Maximum dosage is 35 repetitions. This exercise should be done at a rapid cadence.

(5) Exercise 5: Isometric Bar Lift. Adjust the lower bar so that it is slightly higher than the beltline. Placing the feet on the footplates at the base of the frame, grasp the lower bar so that the hands are spread shoulder-width apart. Assume a crouched position and lift with maximum effort using the arms, back and legs.(See fig. 3-40, E.)Starting dosage is 4 repetitions of a stress time of 5 seconds followed by a 5-second rest prior to the next repetition. The Marine obtains progression by lengthening stress periods to 6 and later 7 seconds. Do not increase the number of repetitions.

(6) Exercise 6: Knee Lift. Adjust the upper bar to the same height used in Exercise 2.Adjust the lower bar so that is stops rearward movement of the hips when the "dead hanging" position is assumed.(See fig. 3-40, F.) Keeping the arms extended, flex the legs and raise the knees as high as possible. Hold this position for 5 seconds, then return to the starting position. After 2 seconds in the starting position, raise the knees again. Each return to the starting position constitutes one repetition. The dosage is five repetitions. Progression is the same as in Exercise 5.

(7) Exercise 7: Isometric Pull. Adjust the lower horizontal bar to a position where it is slightly higher than the beltline. Grasp the handles and pull outward.(See fig. 3-40, G.)Apply maximum effort and hold for approximately 5 seconds. Relax for 5 seconds between repetitions; perform four repetitions. Moving the body closer to or farther away from the bar will change the stress from the upper arms to the forearms. Progression is the same as in Exercise 5.

(8) Exercise 8: Isometric Compression. Maintain the position as in exercise 7.(See fig. 3-40, H.)Grasping the handles in the same manner, press in **with** maximum effort and hold for approximately 5 seconds. Relax for 5 seconds between repetitions. Progression is the same as in Exercise 5.

(9) Exercise 9: Isometric Press. Remove the lower horizontal bar. Adjust the upper horizontal bar until it is about 6 inches lower than the extended arms can reach. Stepping on the footplates at the bottom of the frame, grasp the bar with both hands and push up.(See fig. 3-40,1.)Keep both the legs and arms slightly flexed and the back straight. Apply maximum effort for 5 seconds then relax for 5 seconds. Complete four repetitions. Progression is the same as in Exercise 5.

(10) Exercise 10: Pushups. Grasping the foot braces with both hands, assume the front leaning rest position.(See fig. 3-40, J.)Keeping the back and legs straight, lower the body until the chest is lower than the **hands,** then return to the starting position. The Marine should complete the maximum possible number of repetitions.

f. Bench Conditioning, Table II. There is no limit on the maximum number of repetitions attainable in Exercises 3, 4, and 6 of Table II. The only limit imposed is that the entire program of 10 exercises outlinted in either table should not exceed 15 minutes.

(1) Exercise 1: Side-Straddle Hop. This is a two-count warm up exercise done at a moderate cadence. The starting position is the position of attention. On count ONE, jump slightly into the air, swinging the arms out to the sides and up to a vertical position, hands touching.(See fig. 3-41, A.)At the same time, spread the feet wider than shoulder-width apart. On count TWO, using a slight flexing of the knees and ankles, jump slightly into the air and return to the starting position by swinging the arms back down to the sides. Twenty repetitions of this exercise is the standard dosage throughout the program.

(2) Exercise 2: Pull-up. Adjust the horizontal bar so that is is high enough to permit a "dead hanging" position with the feet off the ground. Grasp the bar with both hands, palms facing forward. By flexing the arms, raise the body to a position where the chin is higher than the bar.(See fig. 3-41, B.)Then lower the body to the "dead hanging" position. Repeat as many times as possible.

(3) Exercise 3: Bench situps. Sit on the bench and hook the feet under the foot braces. With the fingers interlocked behind the head, lean back until the head touches the floor.(See fig. 3-41, C.)Return to the starting position. The starting dosage is 15 situps

(4) Exercise 4: Step-Up. Face the bench and step up on it with one foot, bringing the trailing foot up next to the leading foot. Step back down again, leading with the same foot used first in stepping up.(See fig. 3-41, D.)Perlorm half of the total repetitions, then change the sequence of moving the feet to use the other leg in stepping up, and repeat the same amount of exercise. The starting level is a total of 40 step-up. **This exercise should be done at a rapid cadence.**

(5) Exercise 5: Isometric Bar Lift\djust the lower bar so that it is slightly higher than the beltline. Placing the feet on the footplates at the base of the frame, grasp the lower bar so that the hands are spread shoulder-width apart. Assume a crouched position and lift with maximum effort using the arms, back, and legs.(See fig. 3-41, E.)Starting dosage is four repetitions of a stress time of 8 seconds followed by a 5-second rest prior to the next repetition. Progression is obtained by lengthening stress periods to 10 seconds. Do not increase the number of repetitions.

Figure 3-41 is available in the illustration section at the end of the book.

(6)Exercise 6: Leg Lift. Adjust the bars and assume the starting position as shown in figure 3-41, F. Keeping arms and legs extended, raise the legs to a horizontal position and hold in that position for 2 seconds. Then lower the legs slowly to the starting position. Five repetitions is the starting level.

(7) Exercise 7: Isometric Pull. Adjust the lower horizontal bar so that it is slightly higher than the beltline. Grasp the handles and pull outward.(See fig. 3-41, G.)Apply maximum effort and hold for approximately 8 seconds. Relax for 5 seconds between repetitions; perform four repetitions. Moving the body closer to or farther away from the bar will change the stress from the upper arms to the forearms. The Marine obtains progression by lengthening the stress period to 10 seconds.

(8) Exercise 8: Isometric Compression. (See fig. 3-41, H.)Maintain the position as in exercise 7.Grasping the handles in the same manner, press in with maximum effort and hold for approximately 8 seconds. Relax for 5 seconds between repetitions; perform four repetitions. The Marine obtains progression by lengthening the stress period to 10 seconds.

(9) Exercise 9: Isometric Pres. Remove the lower horizontal bar. Adjust the upper horizontal bar until it is about 6 inches lower than the extended arms can reach. Stepping on the footplates at the bottom of the frame, grasp the bar with both hands and push up.(See fig. 3-41,1.)Keep both the legs and arms slightly flexed and the back straight. Apply maximum effort for 8 seconds, then relax for 5 seconds. Complete four repetitions. The Marine obtains progression by lengthening the stress period to 10 seconds.

(10) Exercise 10: Inclined Pushup. Assume the front leaning rest position with the feet on the bench.(See fig. 3-41, J.)Keeping the back and legs straight, lower the body until the nose touches the ground. By extending the arms, raise the body to the starting position. Repeat as many times as possible.

3903. THE 6-12 PLAN

a. Description and objective. The 6-12 plan of physical fitness has been developed to assist in regulating quantity and progression and to provide a convenient set of exercises. This is a basic program and will take 18 weeks to complete if you follow the moderate progression as to the time prescribed for each level of achievement. The time can be shortened as explained below. This plan consists of six basic exercises a day which can be completed in 12 minutes. There are six tables of six exercises each, thus allowing you to progress from table to table. The plan is progressive, fits any age group, contains balance and variety, and applies the principle of overload in a safe and gradual manner. Begin at Table I, Progression Guide, with the number of repetitions as indicated by age.

b. Formation. If just starting an exercise program, do not rush through the first table. Remember, individuals should remain at each level for about a week before moving upward. The time allotment stated for each exercise at the bottom of the tables is a guide; some people may take more and some less time on the individual exercises. At the end of a one-week period, when the individual can comfortable perform the six exercises in 12 minutes, move on to the next level. To a certain degree, the individual must be the judge of his ability to progress form level to level and table to table. If starting with a certain degree of fitness, some of the beginning tables may present little challenge.

c. Starting Level **and Progression.** There are three levels of achievement for each age group, indicated as A, B, and C. Start at the C level for the appropriate age group. At the end of a one-week period, or when the individual can do all exercises at that level within 12 minutes, progress to the B level. At the end of the second week, or when the individual can accomplish that level within 12 minutes, progress to the A level. At the conclusion of the third week or when the individual can achieve the A level within the time limitation, move on to table II.(See Tables I through VI, Progression Guide.)

d. Maintenance Level. Attempt to work through all six tables. If this proves to be too difficult, then maintain exercise at the—

(1) A-level on Table IV, Progression Guide it in the 45 to 49, 50 to 59, or over 60 age group.

(2) A-level on Table V, Progression Guide if in the 17 to 29, 30 to 39, or 40 to 44 age group.

e. Precautions. To achieve the maximum benefit, perform each exercise exactly as specified. Read the descriptions and study the illustrations. Do not slight the movements. Use a sensible approach and follow these four points as they apply before starting or during your exercise program.

(1) If you have the slightest doubt about your ability to participate in this exercise program, consult a physician.

(2) Stop immediately if you notice unusual breathlessness or chest pain while taking part in these exercises. If these conditions persist, consult a physician.

The following Tables of Progression Guides and Exercises are the progression of the 6-12 Plan program.

Table I. Progression Guide is available in the illustration section at the end of the book.

Table I. Exercises.is available in the illustration section at the end of the book.

Table II. Progression Guide.is available in the illustration section at the end of the book.

Table II. Exercises.is available in the illustration section at the end of the book.

Table III. Progression Guide.is available in the illustration section at the end of the book.

Table III. Exercises.is available in the illustration section at the end of the book.

Table IV. Progression Guide.is available in the illustration section at the end of the book.

Table IV. Exercises.is available in the illustration section at the end of the book.

Table V. Progression Guide.is available in the illustration section at the end of the book.

Table V. Exercises.is available in the illustration section at the end of the book.

Table VI. Progression Guide.is available in the illustration section at the end of the book.

Table VI. Exercises.is available in the illustration section at the end of the book.

(3) Unless you have exercised regularly and know yourself to be in good physical condition, start at table I with the C level appropriate to your age.

(4) If you are out of shape, admit that fact to yourself. Hide your pride; after all, you are in the privacy of your own quarters. Set your goal for the longer, steadier pull toward fitness. Resist the urge to pass over the lower numbered tables to find a table that will test your fitness. You are not trying to test, but rather to develop.

3904.WEIGHT TRAINING

Weight or barbell training should not be confused with the more common types of weight lifting used as a competitive sport. Weight lifting is designed to develop specific muscle groups so that the individual is capable of lifting a large amount during a single lift. In contrast, weight training is the systematic development of all the major muscle groups by the use of calisthenics reinforced with weight to provide resistance.

3905.WEIGHT LIFTING

a. Description and Objective. The weight lifting program is progressive and applies the principle of overload in a safe, gradual manner. The exercises of the table can be completed in 15 minutes. The objective of the exercises is to develop strength and muscular endurance, and muscle tone of the five major muscle groups:legs, arms, back, trunk, and shoulder girdle.(See fig. 3-42.)

b. Warmup. A warm up exercise is important to prepare the body for the more vigorous exercises that are to follow. Ten repetitions of the high jumper exercise are excellent for a warm up period. Also, an exercise which will require fast body movement is needed.To provide such exercise, 3 to 5 minutes of rope skipping is recommended to increase the individual's development potential.

c. Formation. Marines should take care in the completion of these weight lifting exercises. They should ensure that their back is straight during the lifting phase of all exercises. When exercises require assuming the standing position with the weight, they should always grasp the weight while in a squatting position and then rise to a standing position.

d. Progression. Each exercise has a starting number of repetitions and specified pounds of weight. After each 4th or 5th day of exercise, the Marine should increase the repetitions by one until he has reached the maximum of 10 repetitions. At this time, the Marine should increase the weight by 5 pounds. Then he should repeat the process again with the initial number or repetitions.

e. Weight Training

(1) Exercise 1:Squat(See fig. 3-42, A.)Starting level—six repetitions, 50 pounds (commonly called the flatfoot deep knee bend). Place the bar upon the shoulders. Stand with feet about 18 inches apart. Keeping the feet flat, lower the body into the low squat position. Stand erect and repeat. Exhale while lowering into the squat position and inhale while standing. This constitutes one repetition.

(2) Exercise 2:Waist Bender(.See fig. 3-42, B.)Starting level-6 repetitions, 40 pounds. Assume the standing position with the bar across the shoulders and the feet a shoulder-width apart .Bend forward at the waist until the upper body is parallel to the ground; return to the starting position. Each return to the upright position constitutes one repetition.

(3) Exercise 3:Curl (See fig. 3 42, C.)Starting level—6 repetitions, 40 pounds. Grasp the barbell with the palms facing to the rear and assume the standing position with the feet a should-width apart. With the barbell held in front of the hips, flex the elbows and lift the weight until the bar touches the upper chest. Lower the barbell back to the hip level position. Inhale deeply with the upward movement and exhale on the downward movement. Each time that the bar touches the chest will constitute one repetition.

(4) Exercise 4:Side Bender(.See fig. 3-42, D.)Starting level—6 repetitions per side, 40 pounds. Assume the standing position, with the bar across the shoulders, with feet a shoulder-width apart. Bend to the left as far as possible and return to the starting position. Repeat six times and then execute the same procedure to the right for six repetitions.

(5) Exercise 5:Standing Press(See fig. 3-42, E.)Starting level—6 repetitions, 45 pounds. Grasp the bar with the palms facing forward and assume the starting position. Curl the weight to the upper chest position. Inhale deeply and press the bar upward to an overhead position. Exhale while lowering the bar to the chest position. Each time that the bar is pressed upward constitutes one repetition.

(6) Exercise 6:Upward Row. .see fig. 3-42, F.)Starting level—6 repetitions, 40 pounds. Grasp the bar, hands close together, palms to the rear, and assume the standing position. Starting with the bar held in front of the hips, flexing the elbows and the shoulder girdle muscles, lift the bar straight up to an overhead position. Inhale deeply while lifting the bar. Exhale while lowering the bar to the hip position. Each time that the bar returns to the hips constitutes one repetition.

(7) Exercise 7:Shoulder Curl. See fig. 3-42, G.)Starting level—6 repetitions, 25 pounds. Grasp the bar, palms down, and assume the standing position. Keeping the elbows locked, curl the bar, pivoting the arms at the shoulders until the bar is in an overhead position and as far to the rear as possible. Return the bar in the same manner to the hip position. Each time that the bar returns to the hip position constitutes one repetition.

Figure 3-42 is available in the illustration section at the end of the book.

3906.ISOMETRIC EXERCISING

a. Description and Objective. Isometric exercising is the application of maximum effort during an exercise period. The isometric principle is to apply force gradually over a 5- to 10- second period until the maximum effort is applied. Relaxation follows for approximately 5 seconds and then force is applied again. The Marine continues this process at the prescribed level for each exercise for a period of 15 minutes or less. The objective of isometric exercises is to create muscle growth. Isometric exercises are the fastest means of creating muscle growth. They are founded on the fact that a muscle will grow only so fast regardless of the type or duration of the activity. The principle of overload—that the muscles develop commensurate with demand—reinforces that fact. However, isometrics will not develop cardiovascular or muscular endurance. Consequently, Marines who choose to use isometrics in their exercise routine must also include running or some other type of aerobic activity.

b. **Area and Equipment** isometric exercises may be designed to be performed with or without equipment.

c. Use With Other Programs. The isometric exercises presented in this section will provide a variety from which to choose. In addition to the exercises contained in this section, the application of isometric force is used in some of the exercises in section II.

d.Door Frame Exercises. The following exercises are designed for use with a standard door frame found in all offices or barracks.(See fig. 3-43.)

(1) Exercise l:Arm Press. Stand in the doorway with the legs straight, knees locked. Using the arm muscles, press hard upward against the top of the door frame. Repeat for three repetitions applying gradual effort to maximum contraction.

(2) Exercise 2:Leg Press. Stand in the doorway with the hands on the top of the door frame, elbows locked .With knees bent, press hard with the leg muscles .Repeat for three repetitions beginning with a gradual effort and increasing to maximum contraction. A low platform may be necessary to reach the top of the door frame and still maintain a bent knee position.

(3) Exercise 3:Side Press. Extend both arms to the side of the doorway. Palms are shoulder high, facing outward. With both arms, press hard against the sides of the door frame. Repeat for three repetitions. Begin gradually and increase to maximum contraction.

(4) Exercise 4:Lateral Raise. Extend both arms to the sides of the doorway, arms down, palms facing inward. With the back of the hands, press hard against the sides of the door frame .Repeat for three repetitions. Begin with a gradual effort and increase to maximum contraction.

Figure 3-43 is available in the illustration section at the end of the book.

(5) Exercise 5:Neck Press?lace the forehead against the door frame, hands clasped behind the back Using the neck muscles, press hard against the door frame. Repeat for three repetitions, then reverse position so that the back of the head is resting on the door frame. Again do three repetitions .Begin gradually with both exercises and increase to maximum contraction.

(6) Exercise 6:Door Pull Stand facing the edge of the open door and grasp the doorknobs. Pull outward with both arms (if doorknobs are not available grasp the edge of the door).Applying outward pressure, move the body toward and away from the door .Repeat the three repetitions. Begin with gradual effort and increase to maximum contraction.

Chapter 4
COMBAT WATER SURVIVAL
Section I. Marine Corps Water Survival Program

4101.MILITARY SWIMMING

Military swimming emphasizes strokes that result in staying power rather than speed. Marines need to swim easily, quietly, and with adequate vision. Strokes should be used that allow the Marine to carry basic combat equipment and to tow or push a wounded buddy while keeping the face out of the water to allow breathing. The best strokes to use are the side stroke and the breast stroke. Marines must be proficient in floating, drown-proofing, and in using their basic combat equipment to make flotation devices. They must be drilled on using the Kapok and Mae West buoyancy compensators and on emergency egress from helicopters, landing craft, and amphibious assault vehicles.

4102.PROGRAM DEVELOPMENT

Commanders of the Landing Force Training Commands (Atlantic/Pacific) are tasked with developing and maintaining the individual water survival and swimming training program. They are responsible for training all Marine Corps water safety and survival instructors and integrating this training in to their combat readiness training program. All Marines are deployable and as soldiers of the sea must constantly practice the skills necessary to achieve mission accomplishment. Commanders must be imaginative in program development, integrating basic skills development which enhances the total mission accomplishment. An example is teaching waterproofing of the Marine's backpack so that it can be used as a flotation device in streams or rivers. Once learned, this skill would be practiced first in a pool and then in a stream or a river. Routine inspections during deployments would ensure that backpacks could be used as flotation devices.

4103.QUAL STDS AND TRAINING GUIDELINES

The inherent nature of Marine Corps operations and training requires that Marines achieve an ability to survive in water. Water survival and swimming training is designed to reduce fear of water, instill self-confidence, and develop a Marine's ability to survive in water. As a minimum, every Marine should be qualified as swimmer, third-class (S3).This qualification should be met during recruit training or Officer Candidate School. Once qualified, a Marine need not be requalified unless requalification at a specific level (S3 or higher) is required by unit mission, military occupational specialty or duty assignment, or for other reasons determined by the commander. The following classifications and standards apply to Marine Corps water survival and swimming training. Abbreviations will be used for service, medical, and training record entries citing MCO 1510.29A, Individual Water Survival and Swimming Training, as authority.

a. Qualification standards and test procedures are described in paragraph 4104.

b. Swimmer classification and abbreviations are:

(1) Unqualified - UQ.

(2) Swimmer, third-class (S3).Minimum aquatic skill level for all Marines.

(3) Swimmer, second-class (S2)-

(a) Minimum standard for all naval aviators, naval flight officers, and aircrewmen with written waiver from the commanding officer.

(b) Requirements for military occupational specialty 1803 (assault amphibian vehicle officer), 1833 (assault amphibian vehicle crewman), 1302 (engineer officer), 1371 (combat engineer), 1381 (shore party specialist), and 1379 (engineer operations chief).

(4) Swimmer, first-class (S1)-

(a) Naval aviator, naval flight officer, and aircrewman qualification requirement.

(b) Requirements for military occupational specialty 0321 (reconnaissance man), 8652 (reconnaissance man, parachute jump qualified), 8653 (reconnaissance man, self-contained underwater breathing apparatus qualified), 8654 (reconnaissance man, parachute/self-contained underwater breathing apparatus qualified), 9952 (self-contained underwater breathing apparatus Marine (officer/enlisted)), and 9953 (parachutist/self-contained underwater breathing apparatus Marine (officer/enlisted)).

(5) Water Survival Qualified (WSQ).Ultimate water survival/swimming goal for Marines.

c.Instructor and assistant instructor classifications are-

(1) **American Red Cross Advanced Lifesaver.** Individuals who are currently certified as American Red Cross advanced lifesavers may assist qualified instructors during Marine Corps water survival/swimming training and testing. Certification is valid for 3 years.

(2) American Red Cross Water Safety Instruction. Individuals who are currently certified as American Red Cross water safety instructors may conduct swimming training and qualify individuals as 3d, 2d, and 1st class swimmers. Certification is valid for 2 years.

(3) United States Marine Corps Water Safety and Survival Instruction. Those personnel certified as water safety and survival instructors by one of the landing force training commands may conduct Marine Corps water survival and swimming training and American Red Cross swimming and water safety training. These instructors may also participate as assistant instructors during the conduct of landing force training command-sponsored water safety and survival instructor courses. Certification is valid for2 years.

(4) American Red Cross Water Safety Instructor Traine. Individuals qualified who have been certified as instructor trainers by the American Red Cross may train American Red Cross water safety and basic swimming instructors. Certification is valid for 2 years.

(5) United States Marine Corps Water Safety and Survival Instructor Traine. Individuals qualified who are certified as instructor trainers through special training courses conducted jointly by the American Red Cross and the Marine Corps, under the direction of the Commandant of the Marine Corps, may train water safety and survival instructors and American Red Cross water safety and basic swimming instructors. Certification is valid for 2 years.

d. Training Guidelines Instructors engaged in training Marines as swimmers will use the current edition of the American Red Cross swimming and water safety courses instructors' manuals as the authoritative textbooks for swimming procedures, strokes, breaks, and holds.

4104.WATER SURV/SWIM QUAL STD AND TEST PROCS

a. The following qualification standards and test procedures are applicable to Marine Corps water survival and swimming training. The utility uniform without cover, boots/oxfords or socks will be worn for all swimming tests through swimmer, first class. The utility uniform with boots/oxfords will be worn as prescribed for the water survival qualification.

(1) Swimmer, Third-Class (S3) To qualify as a swimmer, third-class, a Marine must enter the water feet first from a minimum height of 5 feet and remain afloat for 5 minutes. During this time the Marine must swim 50 yards using any stroke or combination of strokes. This test should be taken by all Marines as early as possible in their initial training. Those unable to pass this test will be classified as UQ and should be given instruction in fundamental swimming skills. Those who barely meet requirements are swimmers who need help and should receive additional instruction in fundamental swimming skills, whenever possible.

(2) Swimmer, Second-Class (S2.) To qualify, a Marine must enter the water feet first from a minimum height of 10 feet and remain afloat for 10 minutes. The Marine must swim 100 yards and use the three basic survival swimming strokes (side stroke, backstroke, and breast stroke) for a minimum distance of 25 yards each.

(3) Swimmer, First-Class (S1)A prerequisite to qualification as swimmer, first-class, is successful completion of the test for swimmer, second-class. To qualify as a swimmer, first-class, a Marine must be able to do each of the following:

(a) Approach a person of approximately the same size while in the water, demonstrate a release, get that person in a carry position, and tow the "distressed swimmer" 25 yards.

(b) Enter water feet first and immediately swim underwater for 25 yards. Swimmer is to break the surface twice for breathing during this distance at intervals of approximately 25 feet.

(c) Remove trousers in water, inflate for support, and remain motionless for a minimum of 1 minute.

(d) Swim 220 yards using any survival swimming stroke or any combination of survival strokes.

(4) Water Survival Qualification (WSQ)A prerequisite to water survival qualification is successful completion of the test for swimmer, first-class. To be water survival-qualified, a Mannemust:

(a) Enter water by jumping from a height of 10 feet.

(b) Stay afloat in full utility uniform (less cover) for 1 hour. Boots/oxfords will be removed after 5 minutes in the water but will be retained by the swimmer.

(c) Traverse 75 yards in deep water in full utility uniform (less cover) with boots/oxfords.

b. All water survival and swimming training will include instruction in the following:

(1) Employment of safety lines or other comparable expedients to assist in swift stream/river crossings.

(2) Adverse physiological effects caused by cold water (hypothermia) and the precautionary measures to be taken prior to exposure to such an environment.

(3) Employment of a standard and expedient flotation devices in a water survival situation.

(4) Removal of individual field equipment after unscheduled water entry. Entry will be made from a minimum height of 10 feet with a complete set of field equipment, properly packed as an expedient flotation device and waterproofed. The ALICE pack will then be removed, the rifle laid across the top of the pack, the helmet remaining or put back on the individual's head, and the individual will traverse 25 yards to the side of the pool or water's edge.

c. The additional water survival qualification (flight physiology training) and requalification required of personnel assigned to flight status will be conducted in accordance with OPNAVINST 3710.7_,Promulgation of NATOPS General Flight and Operating Instructions. Swimming qualification (first-class, SI) is a prerequisite to water survival (flight physiology) training and is a one-time requirement.

Section II. Water Sports and Competitive Activities

4201.GENERAL

The commander, when developing the combat readiness program, needs to look carefully at the mission. If the mission requires Marines who must perform mission-essential water-related activities of long duration, he may want to look to water sports/competitive activities as a method of developing endurance. Endurance and confidence can only be developed by extended time in the water. While defined training can accomplish this goal, the goal can be enhanced through training which is integrated with the physical conditioning program. This approach will provide variety and enthusiasm. Two activities which will accomplish this goal are water relays and water polo. The important consideration is that the unit achieve total participation by all members of the unit.

4202.WATER RELAYS

Development of water relays will generally conform to the guidelines in chapter 5, section IV. When contemplating relays, the commander should review chapter 5 and adapt the principles to the size of the unit and the facility to be used.

4203.WATER POLO

The official rules provided in this paragraph will need to be adapted to the unit's unique situation, its size, and the size of the facility. Commanders should make every effort to have all members of the unit participate.

a. NCAA Rules This style of play is based on two popular sports: swimming and basketball. Playing is done on the surface of the water by teams of seven players each. The size of the playing area is 25 yards by 14 yards, which is the size of the typical high school or college indoor pool. The deeper the water, the better. If the entire playing area is deep-6 feet or more—this is ideal, although most indoor pools have a shallow end. At each end of the pool is a goal. In deep water, the goal is 10 feet across and 3 feet high. In shallow water, where the goalie can stand on and jump from the bottom, the goal is 10 feet across and 5 feet high. One of the seven players on each team is the goalie. He is the only player on the team who can stand on or jump from the bottom (if it is shallow enough to permit this), catch and pass the ball with both hands at the same time, or hit the ball with a clenched fist. However, the goalie may not swim across or pass the ball across the middle of the pool.

b.Equipment. The goalie on the visiting team must wear a white cap with the number 1 clearly marked on it; his teammates, who are designated as guards and forwards, must wear white caps numbered 2 through 7, with the substitutes wearing white caps numbered 8 through 21.The goalie on the home team must wear a dark-colored cap with the number 22 clearly marked on it; his teammates designated as guards and forwards, must wear similar dark-colored caps numbered 29 through 42.The guards and forwards on each team may swim freely up and down the pool as they see fit, interchanging positions as often as they wish. They may not stand on or jump from the bottom, touch the ball with more than one hand at a time when catching or passing or shooting it, or enter inside the opposing team's 2-yard line unless preceded by the ball. The ball is similar to a soccer ball, except that it is yellow and covered with a waterproofing substance that makes it easier to handle with one hand.

c. Playing Time. In intercollegiate competition the game consists of four 7-minule quarters, the teams changing ends after every period of play. In high school competition the game consists of four-,five-, or six-minute quarters, depending on the ages of the participants. Here again the teams change ends after every period. There is a very brief interval between each quarter. At the start of each period the teams line up at their respective ends, and with ablast of the whistle, the head referee throws the ball into the middle of the pool. The fastest swimmers on each team then race to gain possession.

d. Officials. The head referee walks along one side of the pool watching for infractions. He is aided by the assistant referee, who patrols the opposite side of the pool. The referees are both equipped with whistles and two official's flags. These flags should be 12 inches square and should correspond in color to the caps worn by the respective teams. Whenever a referee sees an infraction, he blows his whistle and signals with his flag indicating that the player nearest to the ball with the appropriate colored flag will be given possession of the ball. Whenever the whistle is blown, all players should react by looking immediately to see which color flag is being waived which will indicate which team should be moving into the defense. Other officials are a timer and a scorer. As in all sports, it is important that the officials be skilled and competent.

e. Technical Fouls. The following are some of the common technical fouls: starting before the referee blows his whistle to open play; holding onto or pushing off from the side of the pool during play; taking hold of the ball underwater when tackled by an opposing player; swimming inside an opponent's 2-yard line unless preceded by the ball; touching the ball with both hands at the same time (goalie excepted); standing, walking on or jumping from the bottom when taking an active part in the game. When a technical foul occurs, thereferee blows the whistle and with his flag awards possession. The player on the team awarded possession nearest the point of infraction then has 5 seconds to put the ball into play by passing to a teammate or by dropping the ball into the water and swimming after it.

f. Personal Fouls. The following are some of the more common personal fouls: committing any of the technical fouls mentioned above for the purpose of scoring or preventing a goal; holding, ducking, pulling, pushing off from, swimming over, or impeding the arm or leg movement of any opponent who is not touching the ball; and splashing water in the face of an opponent. When a personal foul occurs, the referee will follow the same procedure as with a technical foul. While doing this, the referee will also call out clearly and loudly the number of the player who was guilty of the infraction and a personal foul will be marked against that player by the official scorer. When a player accumulates five personal fouls, he fouls out of the game and must be replaced by a substitute.

g. Penalty Shot. A penalty shot can be awarded by either of the officials or the scorer when:

o An offensive player inside the opponents' 4-yard line not touching the ball is held, ducked, pulled-back, kicked, or struck.

o A team has accumulated a total of ten personal fouls. When the former occurs, the head or assistant referee should immediately blow the whistle and by holding a two-flagged stick in a vertical position above his head, signal that a penalty shot has been awarded. When the latter occurs, the scorer should use a buzzer located at the scorer's tabic to signal that one team has accumulated ten personal fouls, thereby entitling the other team to a penalty shot. A penalty shot is taken from the 4-yard line in front of the goal. All players except the defending goalie must leave the 4-yard line until the shot is taken. No player can be within 1 yard of the shooter. After ascertaining that the shooting player is on the 4-yard line and the goalie is on The goal line, the referee will ask the shooter to lift up the ball. When he docs, the referee will give a sharp blast from the whistle. At the whistle, the shooter must shoot without delay and without any faking at the goal. The goalie may try to block the shot. If he does, or the shot is otherwise missed, it is immediately in play and action continues.

h. When a Goal is Scored. When a goal is scored, either from a shot taken by a player out in the field or by penalty shot, the defending goalie must then pass the ball to the nearest referee. The referee will then pass the ball back to the goalie, who puts it into play by a pass to one of his teammates as soon as possible.

Section III. The Battle Swimming Test

4301.GENERAL

The battle fitness test is the evaluation contained in the Guidance for Basic Warrior Training (BWT) as part of the Basic Warrior Training Concept Plan.

4302.THE BATTLE SWIMMING TEST

The battle swim test consist of the following:

a. Be able to climb a three meter high board, while wearing a field uniform, carrying a M-16, web gear, and four canteens. Jump off blindfolded, recover and swim 100 meters with the M-16.

b. Repeat the same process, without the blindfold, and upon entering the water a second time, recover and retrieve all submerged gear.

c. Remain afloat and tread water for five minutes.

d. Demonstrate a knowledge of how to cross a stream using field expedient techniques and a single strand rope bridge.

e. Successfully complete helicopter egress training.

Chapter 5
COMPETITIVE CONDITIONING ACTIVITIES
Section I. Organization of Competitive Activities

5101.LEADERSHIP OF COMPETITIVE ACTIVITIES

a. Place in the Program. Competitive conditioning activities consist of dual combatives, relays, team contests, and team sports in which individuals or teams compete against an opponent to win. Competition is one of the best ways of maintaining interest in the physical training program during the sustaining stage. Organized competition provides enjoyable, vigorous physical activity that has proved to be one of the best supplements to conditioning drill activities. The benefits of competition are the development of aggressiveness, unit pride and identity, teamwork, and the will to win. These activities help develop cardiovascular and muscular endurance, strength, and coordination. Scheduling competitive activities in an orderly and progressive manner is desirable. The progression should be from relays to dual combatives, to team contests, and finally to team athletics. These activities should be conducted as part of the program after a basic level of conditioning has been developed, usually during the slow-improvement stage of conditioning. Muscles and joints should be strengthened by preconditioning to withstand the strain placed upon them by sudden stops and turns, body contact, bearing of weight, and falls. Competitive activities, however, should not be allowed to dominate the physical training program.

b. Area and Equipment. Some of the competitive activities included in this chapter require specific types of areas and equipment; others do not. The area requirement can usually be satisfied on available training fields. When items of equipment are required or specific courts or field layouts are to be marked off, such information is included.

c. Leadership. The principal factor for success as a leader of competitive activity is an energetic, dynamic, enthusiastic approach. The leader's attitude is reflected by the group, so the commander must carry on the activity in a snappy and vigorous manner. Confidence on the part of the commander will create an impression of decisiveness and certainty. Confidence grows out of experience and a thorough knowledge of the activity. Mastery of subject is the first step in developing confidence, assurance, and poise.

(1) The following suggestions are offered for leaders of competitive combatives, contests, and sports:

(a) Get the activity underway quickly by selecting and teaching only the essentials.

(b) Use rules to add to the enjoyment of the activity and do not allow them to interfere with the spirit of competition.

(c) Stop the activity before interest begins to lag.

(d) In team contests, clearly distinguish sides by attempting to maintain unit identity.

(e) Always insist on fair play, enforcing the rules impartially.

(2) The following procedure is recommended for presenting a competitive activity.

(a) Name the activity.

(b) Briefly explain the objective of the activity and give only the pertinent rules.
(c) Have a demonstration at slow speed and answer questions.
(d) Organize groups into teams and appoint captains.
(e) Arrange teams in the proper starting positions.
(f) Conduct the activity.

d. Competitive Units. Units for competition should be the same organizations in which Marines train: battalions, companies, platoons. In most situations, the unit is the squad.

e. Provisions for Instruction. One of the most effective methods of maintaining interest and participation in competitive activities is to provide instruction in activities with which most Marines are unacquainted. Such instruction can be conducted during regular physical training periods. Careful planning is required to keep all individuals continuously engaged in vigorous activity. The materials in the following sections should be used as a guide for instructional purposes.

f. Officiating and Control. Every effort should be made to provide good officials for all competitive activities. Poor officials quickly cause dissatisfaction among participants in team activities and create a situation which can turn an organized activity into a brawl. Each company should have or develop several competent and qualified officials available for games on company and platoon levels.

5102.MILITARY FIELD MEETS

When units reach the latter part of the slow-improvement and sustaining stage of training, interest in the program may lag. A change in course content can arouse the desire to participate. An event which does not require a high degree of skill yet demands strenuous activity is ideal for the military field meet. This is a series of team contests conducted on a station-to-station basis during a given period of time. Team contests carried on simultaneously provide essential training, vigorous exercise, and stimulating competition in an atmosphere that is enjoyable for all. Because of the healthy rivalry that a field meet arouses, it is an excellent form of inter-unit competition.

a. Objective and Advantages. The objective of the military field meet is to provide activity for everyone in the participating units. Activities are chosen that will develop aggressiveness, teamwork, a will to win, and competitive spirit, and that will stimulate interest and build esprit de corps. The military field meet can **be** included as part of the physical training program or as part of the off-duty recreational program. It is a form of contest that can be conducted in nearly all circumstances because it can be easily modified, requires a minimum amount of equipment, and can be readily organized.

b. Level of Competition. The military field meet is flexible. It can be adjusted to large or small groups. A company-size unit is the most desirable, but it may also be administered within a larger unit. If it is conducted within a company, the participating units will be squads. If it is held within a larger unit, platoons compose the teams. A larger-size unit requires more extensive organization and administration.

c. Selection of Events. In organizing a military field meet, select events that are simple and easy to administer. All rules and regulations should be clearly understood by everyone. No event should require previous practice. In selecting the events, consider the interest and capabilities of the Marines and available equipment and facilities. Select events which will require that all members of the units participate. Events should not be dominated by the athletes; rather, MOS-related events which have significance in producing skills and combat readiness should be used.

d.Equipment and Facilities. The site must be large enough to permit events to be grouped about a central control point. Use available facilities such as parade decks, Softball fields, or tracks. A public address system is desirable at the control point for the initial orientation of teams and for subsequent announcements of time lapses, cumulative scores, and final standings. A tally board is necessary so that all units can constantly monitor how the events are progressing.

e.Administrators. Efficient administration of the military field meet depends on the referees, judges, and scorers. For these positions, choose individuals within the units who have had athletic or officiating experience. Prior to the day of the event, all administrators should be briefed and assigned a specific task in order that they may become familiar with the rules and organization of the contests they will conduct. The following officials should be available:

(1) A primary instructor or supervisor who is in charge of the control center and who is responsible for the successful operation of the athletic carnival. The supervisor must have an assistant to act as a timer and scorer.

(2) One assistant instructor in charge of each event. This person should be-

(a) Familiar with the rules of the event.

(b) Effective as a leader to ensure proper supervision and control over the participating teams.

(c) Enthusiastic to provide proper motivation.

(d) Self-confident of the ability to judge infractions of the rules. The assistant instructor must be fair in judgment and penalize without hesitation when infractions occur.

(3) Runners between event stations and the control point. These individuals collect and deliver scores.

f. Team Organization for Competition. The size of the teams is determined by the level on which the field meet is organized. Maintaining the integrity of the unit promotes esprit de corps, but this does not preclude grouping two squads into one team.

g.Conduct of Events

(1) The assistant instructor at each station takes charge of the group and gives a brief explanation of the major rules of the event. The assistant instructor speaks clearly and distinctly from a position to be seen and heard by all. A short demonstration is desirable if it will help clarify the event.

(2) The assistant instructor should make certain that teams can be clearly distinguished by the use of, for example, T-shirts and fatigue jackets, or caps and no cap, or colored jerseys. The assistant instructor should teach a whistle response (teams stop play immediately upon hearing whistle), get the event started as quickly as possible, and make any necessary corrections as the contest progresses. The rale of good officiating is to use a minimum of calls, yet maintain control of the contest. Penalize when necessary, but refrain from disqualifying contestants or teams.

(3) Keep the activity moving as rapidly as possible. When the central control point sounds the whistle to stop the play, all competition ends immediately. The assistant instructor then assembles the group, forwards the team scores to the central point, and, upon a signal from the central control point, rotates the teams to their next station. It is essential that rotation and orientation of teams be quick and orderly to reduce time spent between contests.

(4) Upon completion of the final event, assistant instructors move their teams to the central control point for the announcement of winning teams, presentation of awards (if any), and final critique.

h. Scoring System. The system for determining the winner of the military field meet should be simple and efficient. At the completion of competition, the scorer totals the points that each team has scored in all contests. The scorer then subtracts the number of points scored against a team (penalty points for nonparticipation,etc.) from the total. The resulting scores are placed in a column with the highest score at the top and the lowest at the bottom. The team with the highest total is the winner. This type of scoring system encourages a team to ensure total unit participation.

Section II.Combatives

5201.DESCRIPTTON AND OBJECTIVE

Combatives are strenuous, short competitive contests in which two individuals attempt to overcome each other in a bout of skill and strength. These contests help to develop Marines' resourcefulness, confidence, strength, agility, coordination, and will to win. Any level ground area can be used. However, extremely hard ground should be avoided as some of the combatives require ground contact. A whistle is needed to control the bouts during the competition since voice commands may go unheeded. The extended rectangular formation is used for dual combatives. Combatives are conducted on an informal basis. Marines are allowed to remain at ease between activities and are allowed to brush themselves off after being on the ground. To get the most out of combatives, individuals must be urged to overcome their opponents as quickly as possible as would be required in combat.

a. Benefits for Marine Training. The possibility of close contact with an enemy in combat faces Marines at all times. They must be trained to react aggressively and violently. Combatives may be used as an introduction to such hand-to-hand contact and should be followed by hand-to-hand combat training. Marines enjoy competition, and this type of activity is a welcome change from formal conditioning activities.

b. Instructor Responsibility

(1) The instructor tells the individuals that all combatives begin and end on the whistle signal. The instructor demonstrates each activity before having it performed, explaining it in simple terms.

(2) After stopping one activity, the instructor gets everyone in place for the next.

(3) The instructor must closely supervise combatives to insure that contestants do not use unfair or unsportsmanlike tactics. To avoid unnecessary injury, instructors must see that the bouts are closely controlled and opponents equally paired. Adjustments should be made in apparent cases of mismatched abilities.

5202.COMBATIVES TABLES

There are three tables of combatives, each of which can be completed in 15 minutes. These tables become progressively more difficult from lower to higher numbers.

a.Combatives Table I

(1) Open-Hand Slap Boxing(See fig. 5-1, A. individuals assume a boxer's stance, palms open, fingers extended and joined. Each contestant tries to slap the opponent about the head and upper body with the open hand. This is a good warm- up activity.

(2) Wrist Tug-of-War(See fig. 5-1, B.)Two individuals sit on the ground with the soles of their feet in contact. Each grasps the opponent's wrists so that the hands are directly over the feet. At the whistle, each individual tries to pull the opponent from sitting to standing position.

(3) Arm-Lock Wrestling.(See fig. 5-1, C.)Two individuals sit back-to-back with legs spread and arms locked at the elbows. Each right arm is inside the opponent's left. Each tries to force the opponent's left arm or shoulder to the ground. The individual who first wins three bouts is the winner.

(4) Bulling(See fig. 5-1, D.)Two individuals assume the Westmoreland wrestling hold, each grasping the opponent's neck with the right hand and the opponent's right elbow with the left hand. Each tries to force the other to move one foot by pushing, pulling, or otherwise manipulating. The individual who first wins two boutsis the winner.

(5) Indian Wrestling(See fig. 5-1, E.)Two individuals lie on the ground, side by side, with their heads in opposite directions. They link right elbows. On the instructor's signal or by mutual agreement, each individual raises the right leg approximately straight and far enough to engage the opponent's heel. To start the contest, each person usually raises the leg three times rhythmically and on the third time engages the opponent's heel. Each tries to roll the other over backward. The right leg is used for three bouts, then the left leg for three bouts.

Figure 5-1 is available in the illustration section at the end of the book.

b. Combatives Table II

(1) Wrist Bending.(See fig. 5-2, A.) Opponents pair off, face each other, raise their arms forward, and, with palms forward, interlock their fingers. At the starting signal, each individual attempts to bend the opponent's wrist. The arms are kept up and forward and are not allowed to swing around and down to the sides. The individual who first wins two bouts is the winner.

(2) Back-to-Back Push.(See fig. 5-2, B.)Two individuals stand back-to-back with elbows locked. Each right arm is inside the opponent's left arm. At the starting signal, each pushes backward, trying to move the opponent. Opponents are not allowed to lift and carry each other; only pushing is permitted. The one who pushes the opponent the farthest wins the bout. The individual who first wins two bouts is the winner.

(3) Hop and Pull Hand.(See fig. 5-2, C.) The individuals are matched in pairs. Each grasps the opponent's right hand and, hopping on the right foot, attempts to pull the opponent off balance. Contestants automatically lose if they touch their free hand or their lifted foot to the ground. For successive bouts, they alternate hands and feet.

(4) Westmoreland Wrestling. (See fig. 5-2, D.)Each contestant grasps the back of the opponent's neck with the right hand and the opponent's right elbow with the left hand. In this position, each attempts to pull, push, or force the opponent to touch the ground with any part of the body other than the feet. The individual who first wins two bouts is the winner.

(5) Crab Fight.(See fig. 5-2, E.)Two individuals sit on the ground facing in opposite directions with their hands on the ground behind them. At the whistle, they raise their hips and push with their shoulders and bodies, each trying to make the other's hips touch the ground. The individual who first wins two bouts is the winner.

Figure 5-2 is available in the illustration section at the end of the book.

c. Combatives Table III

(1) Hand Wrestling.(See fig. 5-3, A.) Two individuals stand facing each other. Their right feet are forward and braced side by side. They grasp right hands for the first bout, left hands for the second bout. Each pulls, pushes, makes sideward movements, and otherwise maneuvers to force the opponent to move one or both feet from the original position. The contestant who first wins two bouts is the winner.

(2) Back-to-Back Tug.(See fig, 5-3, B.) Two individuals stand back-to-back with both arms linked at the elbows. Each has the right arm inside the opponent's left arm. At the starting signal, each attempts to pull the opponent. Lifting and carrying are permitted. The contestants must maintain their original direction and keep their arms linked. After a predetermined time, the player pulled or carried the farthest is the loser.

(3) Wrestling to Lift off Feet.(See fig. 5-3, C.) Contestants face each other. Each places the right arm under the left arm of the opponent and around the body. The left arm is over the opponent's right shoulder. Each tries to lift the other off the ground. The individual who first wins two bouts is the winner.

(4) Arm Pull Between Legs.(See fig. 5-3, D.) Two individuals are paired off, back-to-back. Each bends forward and, extending the right arm between the legs, grasps the opponent's right wrist. At the starting signal, each person attempts to pull the opponent After a predetermined time, the player who has pulled the opponent the farthest is the winner of the bout. The one who first wins two bouts is the winner. Repeat with the left hand and then both hands.

(5) Rooster Fight.(See fig.5-3, E.) Each contestant grasps the left foot with the right hand from behind, and the right arm with left hand. Each hops on the right foot, and, by shoulder-butting the opponent or by feinting and sudden evasions, forces the opponent to let go of the foot or arm. The contestant who first wins two bouts is winner.

Figure 5-3 is available in the illustration section at the end of the book.

Section III .Relays

5301. DESCRIPTION AND OBJECTIVE

Relays are races in which each member of a team runs one leg of the race. The team effort decides the winner. Relays provide stimulating competition and contribute to the conditioning of personnel. They also develop aggressiveness, team spirit, and the will to win. Relays should be dispersed throughout the program for short periods of time to provide a change of activity.

a. Team Organization. Relays are conducted most efficiently in platoon-size groups. Teams of equal size must be organized. Competitive spirit is encouraged and teams are organized better by basing teams on units such as squads, crews, or sections. Team captains should be designated. Extra Marines may be used as officials. The number of individuals on a relay team should be limited to squad size. If larger teams are used, the runners will spend too much time awaiting their turns and too little time actually participating. Two to six teams are ideal for relay competition. It is difficult to keep track of winners when more teams compete.

b. Administration of Relays

(1) The time spent on any one relay should be relatively short. If one team achieves a substantial lead in a long relay, the competitive spirit and enthusiastic participation of the other teams may decrease. Several short relays are generally better than one long relay.

(2) To maintain competitive spirit throughout a number of relays, determine the teams that win, place, and show in each relay and their total points for all relays. This can be done by awarding points to all teams on the basis of position at the finish of each relay. The team with the greatest number of points is the winner of the entire set of relays.

(3) Difficulties commonly encountered in conducting relays may be avoided by the following procedures:

(a) The last player in a relay race should be conspicuously identified. For example, the last runner can tie a handkerchief around the head or arm, take off the shirt, put on a hat or take it off, or use some other means.

(b) Another way to keep track of the progress of the race is to have each player sit or squat as soon as each is finished.

(c) Judges at the starting line can keep the runners from starting too soon.

(d) To prevent contestants from turning before they run the full distance, they should be required to run around a peg, pole, or assistant instructor.

(e) Batons, handkerchiefs, tent pegs, or other objects should be passed from one runner to the next when relays are run on a circular track.

(4) Before a unit's first participation in a relay, inform participants of the rules and scoring system. Violation of the rules should not result in disqualification. Instead, impose point penalties. A point penalty is imposed by subtracting a point from the team total at the conclusion of the relay.

(5) Careful administration will prevent most violations. For efficient conduct of relays, follow this procedure:

(a) Announce the name of the relay.

(b) Form the individuals in relay position.

(c) Briefly explain the relay and the rules for running it.

(d) Demonstrate.

(e) Have a definite finish line, and ensure that all know where it is.

(f) Answer questions.

(g) Conduct the relay.

(h) Determine winner and award points.

5302. RELAY TABLES

a. Events. The following relays are grouped into a table of activities. Each relay table can be completed in 15 minutes. Thus, relays can be used as a sole activity or as a part of a longer period. Each table is planned for a platoon-size group (30-60Marines). Adequate warmup for participants is provided by conducting several repetitions of exercises 1 and 2 of a conditioning drill. The recommended relay tables require an area 40 by 60 meters in size. Each table provides a variety of activity. The tables are progressive in the overload applied and should be scheduled in numerical order although not necessarily on successive days.

b. Relay Table I

(1) **60-Meter Lane Relay** (See fig. 5-4, A.) Each team is assembled in single file behind the starting line. On signal, the first individual of each team runs to the turn-around line 30 meters away, then runs back and touches the next teammate waiting at the starting line. The winning team is the first team to get its last member across the finish line. If an individual starts before being touched by the preceding runner, the team may be penalized.

(2) Wheelbarrow Race.(See fig. 5-4, B.) The players of each team pair off and line up in single file behind the starting line. The first individual walks on the hands while the partner grasps the ankles. They advance to the distance line (25 meters) behind which they exchange positions and return to the starting line. The rear individual must always hold the partner's ankles. After the first pair returns across the starting line, the next pair starts.

Figure 5-4 is available in the illustration section at the end of the book.

(3) Squad Front Relay .(See fig. 5-4, C.) The teams form in a line along the starting line with a 10-foot interval between teams. The members of each team lock elbows so that they are linked together. At the starting signal, the teams run to the distance line (20meters) where the left flank individual acts as a pivot. The team swings around on the pivot and returns to the base line. If a team breaks its links, it must reform before continuing. The first team to completely cross the base line intact is the winner.

(4) Crab-Walk Race.(See fig. 5-4, D.) The players of each team line up in single file. The first person of each team assumes the crab-walk position with feet forward on the starting line, hands on the ground behind it. At the starting signal, contestants move forward to the distance line (10 meters). They touch the line with their feet and then return to the starting line in the reverse position with the head and hands leading. The second person may not start until the first individual touches the finish line.

c.Relay Table II

(1) **100-Meter Lane Relay.**(See fig. 5-5, A.)This relay is conducted exactly as the first relay of table 1 except that the start and turn-around line are 50 meters apart. This relay provides progression in sprinting.

(2) **Frog-Jump Relay.** (See fig. 5-5, B.) Each team lines up in single file. The first individual assumes a squatting position on the starting line. At the starting signal, the contestant progresses to the distance line (15 meters) and back by leaping forward, catching the weight on the hands, and bringing up the legs to the squat position for the next leap forward.

(3) **Simple Relay.** (See fig. 5-5, C.)Each team lines up in single file. Place a marker on the distance line (20 meters) in front of each team. Each team member, behind the first one, grasps the belt of the individual ahead. At the starting signal, each team runs as a unit to the marker, circles it, and returns to the starting line. The first team to completely cross the starting line intact is thewinner.

(4) **Fireman's Carry Relay.**(See fig. 5-5, D.)The players of each team line up in pairs, one behind the other. One individual in each pair carries the partner to the distance line (30 meters), using the fireman's carry. At the distance line, individuals exchange places and return to the starting line. As a variation, the person to be carried lies on the ground. The one carrying lifts the partner to the proper position. This relay may be performed with the other carries described in guerrilla exercises in chapter 3.

d.Relay Table III

(1) **200**-Meter Circle Relay .(See fig. 5-6, A.)A course is laid out in a circular, rectangular,or oval pattern that is 200 meters around. Each team provides one runner on the starting line. On signal, the runner races around the 200-meter track and touches the next teammate, waiting at the starting line, who runs the same course. Each team member runs one lap of the course.

Figure 5-5 is available in the illustration section at the end of the book.

(2) **Bear-and-Crab Race.**(See fig. 5-6, B.) Each team lines up in single file. At the signal to start, the first contestant in each column assumes the bear-walk position (hands on ground in front), walks to the distance line (15 meters), and then runs back to the starting line. The returning contestant touches off the second individual and goes to the rear of the line. The second contestant assumes the crab-walk position (hands on ground behind) and crab-walks with feet leading to the distance line. This contestant also runs back to the starting line and touches off the next individual, who walks bear fashion. The rest of the members of each team alternate in this manner. The relay ends when the first person is back at the head of the line.

(3) **Pilot Relay.**(See fig. 5-6, C.) The players are grouped in threes, arms interlocked at the elbows, and outside players facing backwards. The middle person runs forward; the two outside individuals run backward. They run to the turning point (15meters), where they start back, this time with the middle person running backward and the two outside individuals running forward. The next set of three players starts when the first set crosses the starting line.

(4) **Saddle-Back Relay.**(See fig. 5-6, D.) Mark two parallel lines 15 meters apart. Each team selects a rider. The remaining members of each team count off. The even-numbered players from each team form in single files behind one line and the odd-numbered players from each team form

in single files behind the other line directly across from their teammates. At the starting signal, the rider mounts the back of the first player of the team who carries the rider across the other line. There the rider changes mounts to the first player in the second line without touching the ground. This person carries the rider to the next player waiting in the first line. The relay continues until all of the mounts have carried the rider. A rider who falls off must mount again at the point of the fall. A rider who falls in changing mounts must get back on the original mount before making the change.

e. Relay Table IV

(1) 100-Meter Circle Relay (See fig. 5-7, A.)A course is laid out in either a circular, rectangular, or oval pattern that is 200 meters around. Each team is divided in half with each half positioned at starting lines on opposite sides of the track. Each runner races halfway around the track and touches a teammate who completes the lap. Each runner then waits in file at the halfway point until touched by the runner and then completes the second half of the lap. The first team to return all runners to their original starting line is the winner.

Figure 5-6 is available in the illustration section at the end of the book.

(2) In-and-Out Relay.(See fig. 5-7, B.) Each team lines up in a file with players 2 meters apart. At the starting signal, the first player runs back through the column in a zigzag fashion. He alternates, going to the right of one teammate and to the left of the next. Upon completing the run, the player lines up 2 meters behind the last one in the column. As soon as the first runner has passed the second person, the latter starts to run. This continues until all the players have realigned their original order. The team that finishes first is the winner. It may be desirable to have this relay continue until all individuals have run through their entire team two or three times in succession.

(3) Circle Race.(See fig. 5-7, **C.**) Each team forms a circle and holds hands with all individuals facing out except one who faces in and is the "driver."At the starting signal, the teams race to the distance line (20 meters) and back, keeping the circle intact. All the individuals in the circle must completely cross the distance line. The "driver" gives directions and orders. When the circle breaks, it must be reformed before it can continue. The first team completely over the starting line is the winner.

(4) Horse-and-Rider Relay.(See fig. 5-7, D.)Each team lines up in a single file. At the signal to start, the second player in each column leaps upon the back of the first who carries the rider across the distance line (30 meters). At the distance line, the rider dismounts and runs back to the starting line. There the rider picks up the third individual in the column, and carries this player to the distance line where the first player has remained. This continues until the last person is carried across the distance line.

Figure 5-7 is available in the illustration section at the end of the book.

Section IV. Team Contests and Athletics

5401. DESCRIPTION AND OBJECTIVE

Team contests are competitive activities in which Marines as a team compete with another team to win. They are guided by simple rules and organization. The function of team contests is to provide competition and an opportunity for body contact and to contribute to the development of physical readiness. In competing and working together as a team, individuals develop aggressiveness, the will to win, and teamwork. Team athletics deserve a prominent place in the physical training program because they contribute to increased combat efficiency. Because of the competitive nature of athletics and their natural appeal, individuals take part with enthusiasm. Athletic teams formed at intramural and higher levels are a strong unifying influence and provide one of the best means of developing esprit de corps.

a. Area and Equipment A level training field is sometimes the only area required. Many contests need no equipment. In contests requiring equipment, the need is for standard items such as logs, balls, nets, goals, and similar types of equipment. Specific requirements for area and equipment are listed with each contest description.

b. Progression. Team contests of a strenuous nature should be introduced after a basic period of conditioning has been completed and individuals are in the slow improvement stage of conditioning. Progression can take place from the less active to the more vigorous contests, and then from the non-contact to the contact or combative-type contests.

c. Necessity for Preconditioning Individuals must undergo conditioning prior to participation in athletics. Muscles, organs, joints, and ligaments not accustomed to stress and strain from sudden stops and starts, falls, body contact, rapid turns, prolonged running, and other rigors of athletic competition are subject to injury. Although athletics should not be introduced until players are physically prepared, there is still opportunity to engage in competition through lead-up contests. Individuals learn many of the skills required for athletics while participating in team contests.

d. Benefits. Athletics are beneficial primarily to sustain interest in the program and to maintain an achieved level of physical fitness. Athletics are a supplement, not a substitute for other types of conditioning activities which should continue. All of the desirable traits of physical fitness cannot be developed through athletics, yet their contribution is significant. For athletics to make a proper contribution to physical conditioning, the selected sports must bevigorous. The team contests recommended here are carefully selected for their simplicity, aggressiveness and applicability to squad- teaching teamwork. They are generally strenuous and involve the whole team throughout the entire duration of the contest.

5402. TEAM CONTESTS

a. Pushball

(1) **Players.** 10 to 50 players on a side. (See fig. 5-8.)

(2) Equipment A large pushball 5 to 6 feet in diameter.

(3) Area. A field 240 to 300 feet long, 120 to 150 feet wide. Mark a center line in the middle of the field parallel to the end lines. Mark a line 45 feet on either side of this center line and parallel to it, extending it across the width of the field. Mark another parallel line 15 feet from each end line, extending it across the width of the field.

(4) Game. Four 10-minute quarters are played. Give 2-minute rests between quarters and 5-minute rests between halves. The object of the game is to propel the ball over the opponent's goal line by pushing, rolling, passing, carrying, or any other way except kicking.

(a) The ball is placed on the center line with the opposing captains 3 feet from the ball. The rest of the players are 45 feet from the ball, on their half of the field. On the referee's starting whistle, the captains immediately play the ball with their respective teams coming to their assistance.

(b) At quarter time, the ball remains dead for 2 minutes at the spot where it was when the quarter ended. At half time, the teams exchange goals. The play is then started as it was in the beginning.

Figure 5-8 is available in the illustration section at the end of the book.

(c) Players may use any means of interfering with an opponent's progress except striking and clipping. Clipping is throwing the body across the back of an opponent's leg, while the opponent is running or standing. Legal use of force may be applied to all opponents whether or not they are playing the ball. For striking an opponent, the offender is removed from the game. The team penalty is half the distance to their goal. The penalty for clipping is the same.

(d) When any part of the ball goes out of bounds, it is a dead ball. The teams line up at right angles to the side lines and 3 feet apart at the point where the ball went out. The referee then tosses the ball between the teams.

(e) When, for any reason, the ball remains in one spot for more than 10 seconds, the referee declares the ball dead. The ball is then put into play as it is for an out-of-bounds situation.

(5) Scoring. A goal is scored when the ball, or any part of it, is propelled across the opponent's end line. A goal counts 5 points. The team scoring a goal has the privilege of trying for a point after the goal. To try for this extra point, the ball is placed on the opponent's 5-yard (or 15-foot) line. The teams line up on either side of this line separated by the width of the ball. Before the whistle blows, one player on each team may place hands on the ball. On the referee's signal, the ball is put into play for 1 minute. If any part of the ball is driven across the goal line in this 1-minute period, the offensive team scores 1 point. The defending team may not score during the opponent's try for the extra point.

b. Line Rush

(1) Players. Any number up to 50 on each side. (See fig. 5-9.)

(2) Equipment. None.

(3) Area. A field, 75 by 100 feet.

(4) Game. One team lines up behind one goal line and the other in midfield. On the starting signal, the team standing behind the goal line seeks to cross the goal line at the opposite end of the field within 30 seconds. The team in the center seeks to prevent it by catching and holding the runners. At the end of 30 seconds, the teams change.

(5) Scoring. Count the number of individuals who have crossed the far goal at the end of 30 seconds. After each team has had from three to five tries, the scores are added and the winner declared. A player scores 1 point when any part of the body is across the goal line.

Figure 5-9 is available in the illustration section at the end of the book.

c. Human Tug-of-War

(1) Players. 10 to 20 on a team.(See fig. 5-10.)

(2) Equipment. None.

(3) Area.40 to 60 feet.

(4) Game. Draw a line in the center of the area. Divide the players into two equal teams. Place them in single file on opposite sides of the center line facing each other. Each individual places the arms around the waist of the teammate in front. The two leaders of the opposing teams grasp each other around the waist. On signal, the teams try to pull each other over the center line within 30 seconds.

(5) Scoring. The team pulled across the center line loses. If neither team is pulled over the center line, but one team breaks its file, that team loses the match.

(6) Variation. Use a 3/4- or 1-inch rope. Space the leading players on each team 10 feet apart, each holding one end of the rope. The team pulled across the center line loses.

Figure 5-10 is available in the illustration section at the end of the book.

d. Master of the Ring

(1) Players. Any number. (See fig. 5-11.)

(2) Equipment. None.

(3) Area. A clearly marked circle large enough to contain all the players.

(4) Game. All the players stand inside the circle. At the signal, all players attempt to throw each other out of the circle. All tactics are fair except unnecessary roughness. When any part of the body touches across the line, the player is out and leaves the circle at once. Several officials are needed to spot the players who cross the line.

(5) Scoring The player who remains in the circle when all the others are out is the master of the ring.

(6) Variation 1. The players are divided into two equal teams. Each team is clearly marked. On signal, each team tries to throw the opponents out of the circle. The winning team is the team that eliminates all opponents from the circle.

(7) Variation 2.The players are divided into two equal teams. Each team sends only one individual into the circle. When one person has been forced out of the circle, the losing side sends in only one player. The team which eliminates all opponents is the winner.

(8) Variation 3.A pit, approximately 4-feet deep, is used rather than a circle on level ground. The contest may then be an individual or team activity.

Figure 5-11 is available in the illustration section at the end of the book.

e. Log Pivot Circle. Space teams far enough apart so each team can pivot in a circle without colliding with any other team. Each team holds a log in the bend of the arms in front of the chest. (See fig.5-12.) At the command **CIRCLE RIGHT, MOVE**,the left-flank individual holds the pivot while the log is carried around 360 degrees in a counter- clockwise motion, back to the original position. This movement may also be performed to the left by facing in the other direction and pivoting clockwise(**CIRCLE RIGHT, MOVE**).Other commands may be used such as: **CIRCLE HALF-RIGHT, CIRCLE HALF- LEFT** and so on. The first team to complete the prescribed movement is the winner.

Figure 5-12 is available in the illustration section at the end of the book.

f. Log-Rolling Race. Each team tries to roll its log a measured distance by pushing it with the hands and driving the body forward with the legs. (See fig. 5-13.)The first team to get the entire length of the log across the finish line wins.

Figure 5-13 is available in the illustration section at the end of the book.

g. Prone Pushing Contest.. Two teams lie prone, facing each other with a log between them. (See fig. 5-14.)Both teams place their hands against the log, keeping their arms straight. Then, by driving with the legs, each team attempts to push the other a measured distance to the rear.

Figure 5-14 is available in the illustration section at the end of the book.

h. Shuttle Relay Race. Each team in this race divides into two parts, A and B. Members of part A run 50 yards with a log held under their right arms. (See fig 5-15.)At the distance line, they give the log to their teammates in part B who bring it back to the starting line. The team pair finishing first is the winner.

Figure 5-15 is available in the illustration section at the end of the book.

5403.CROSS-COUNTRY AND DISTANCE RUNNING

a. Value of Running. Long-distance running gives some benefits that cannot be obtained in the same degree from any other sport. It builds powerful leg muscles, increases lung capacity, and develops endurance. For these reasons, cross-country and distance running should be included in the physical training program. These sports require only a few miles of open space which is generally available. Short cross-country runs and middle-distance runs can be used to supplement other activities, particularly team sports or the sports that develop precision or agility rather than endurance. Short cross-country runs can be scheduled once a week, gradually increasing the distance as the physical condition of the runners improves.

b. Cross-Country Runs. A cross-country run is a distance run held on a course laid out along roads, across fields, over hills, through woods, and on any irregular ground. A flat cinder or dirt track is not a suitable surface for cross-country running. Some runs are as long as 6 miles while others are as short as 3 or 4 miles. If cross-country running is being used to supplement other activities, the 3-mile course is long enough for most. Cross-country runs should be used only after individuals reach the sustaining stage of conditioning. This type of running should then be scheduled occasionally to provide variety. Cross-country running has the advantage of allowing mass participation. Interest can be stimulated by putting the runs on a competitive basis.

c. Practice Methods. Conditioning is essential to distance and cross-country running. Championship distance running depends on stamina, which can be developed only through constant training. An individual of only average ability can become an outstanding distance runner by steady and careful training. Hiking is an excellent method for getting into condition before the season opens. Long walks build up leg muscles. During the first month of the season, training should be gradual, starting with short distances, and increasing day by day. At first the legs will become stiff, but the stiffness gradually disappears if running is practiced for a while every day. To prevent strain, it is essential to limber up thoroughly each day before running.

(1) In training a large group, leaders should be stationed at the head and the rear of the column. They should make every effort to keep the runners together. After determining the abilities of the unit in cross-country running, it is advisable to divide them into three groups. The poorest-conditioned group is started first, the best-conditioned group, last. The starting time of the groups should be staggered so that all of them come in at about the same time.

(2) In preliminary training, running is similar to ordinary road work in that it begins with rather slow jogging, alternating with walking. The speed and distance of the run is gradually increased. As conditioning improves, occasional sprints may be introduced. Atfirst, the distance run is from one-half to 1 mile. It gradually increases to 2 or 3 miles. On completing runs, require runners to walk for 3 or 4 minutes before stopping to permit a gradual cooling off.

d. Facilities and Equipment. There should be at least one timer with a stopwatch (preferably three) for timing the runners. A course 3 or 5 miles long should be measured and marked by one of the three following methods:

(1) Fasten directional arrows to the top of tall posts and place them at every point where the course turns. Such signs should also be placed at every other point where there may be doubt as to the direction of travel.

(2) Place a lime line on the ground over the entire course.

(3) Place flags, clearly visible to the runners.

- **(a)** A red flag indicates a left turn.
- **(b)** A white flag indicates a right turn.
- **(c)** A blue flag indicates the course is straight ahead.

e. Rules/Scoring. A cross-country team consists of seven individuals, unless otherwise agreed. In dual meets, a maximum of 12 on one team may be entered, but 7 or fewer are scored. First place scores 1 point, second place 2, third place 3, and so on. All who finish the course are ranked and tallied in this manner. The team score is then determined by totaling the points scored by the first five of each team to finish. The team scoring the least number of points is the winner. If fewer than five (or the number determined prior to the race) finish, the places of all members of that team are disregarded. If two or more teams score the same number of points, the event is called a tie.

5404.TEAM ATHLETICS

Team athletics add variety to the combat readiness program. At no time should they be allowed to replace intramural activities or to provide practice time for intramural or base teams. If team athletics are integrated into the program, the goal should be total participation and competition. Therefore, the games selected can be played by the skilled or the unskilled. They require limited equipment, yet will teach competitiveness, aggressiveness, and teamwork while they improve conditioning.

a.Soccer. Soccer is one of the best athletic activities for developing endurance, agility, leg strength, and great skill fulness in using the legs. The most popular sport in the world, soccer is the national game of many European, Asian, and Central and South American countries. In recent years, it has become popular in the United States. A soccer ball is the only equipment needed for the game, and Marines can learn to play it easily. Players do not need much skill to participate, but the amount of skill they can develop is almost unlimited.

(1) Place in the Program. Soccer should be introduced during the latter part of the slow improvement stage of physical training. It can be used as a competitive activity in the sustaining stage. It is primarily a spring or fall sport. Any level field is suitable for competition. The boundaries for the soccer field are similar to the dimensions for a football field. (See fig. 5-16.) Goal posts are essential, but they are easily constructed and usually temporary so that they may be removed when not in use.

Figure 5-16 is available in the illustration section at the end of the book.

(2) Basic Skills

(a) Passing. Passing with the feet is the primary means of moving the ball. Short passes are easier to control. Emphasis should be continually placed on the skill of passing.

(b) Dribbling. The ball is dribbled by a series of kicks with the inside or outside of the foot. Do not kick with the toe. Keep the head over the ball when kicking and propel it only a short distance at a time. Keep it close to the feet. If a player lets a ball move far from the feet, an opposing player can easily take it away.

(c) Instep Kicking. The instep kick is the basic soccer kick. The toe does not come in contact with the ball. The toe is pointed downward, and the instep (the area under the shoe laces) is applied to the ball with a vigorous snap from the knee.

(d) Inside-of-the-Foot Kicking. The ball is kicked with the inside of the foot, and the leg is swung from the hip. The toe is turned outward, and the sole of the foot is parallel with the ground as the foot strikes the ball. This kick is used for short passes and for dribbling.

(e) Foot Trapping. The foot trap is the method of stopping the ball by trapping it between the ground and the foot. Place the sole of the foot on top of the ball at the instant it touches the ground, keeping the foot relaxed. This is an effective way to stop a fast-moving ball.

(f) Shin Trapping. .The shin trap is a method of stopping the ball with the shins. Stand just forward of the spot where the ball should strike the ground and allow it to strike the shins in flight or on the bounce. Use either one or both legs from the knee down so as to absorb the bounce.

(g) Body Trapping. .Intercept the ball with any part of the upper body except the arms and hands. Keep the body relaxed and inclined toward the ball. To keep the ball from bouncing, move backwards from it as it strikes the body. This drops the ball at the feet in position for dribbling or passing.

(h) Heading. Heading is a technique for changing the direction of the flight of a ball by butting it with the head. Tense the neck muscles and jump up to meet the ball. Butt the ball with the forehead at about the hairline to reverse its direction. Use the side of the head to deflect it to the side.

(3) Offensive and Defensive Positions. Forwards usually play on the offensive half of the field. Fullbacks usually play on the defensive half of the field. Halfbacks are the backbone of the team. They move forward on the offense and back on defense. The goalkeeper almost always remains within a few feet of the goal.

(4) Abridged Rules

(a) A soccer team is composed of 11 players. (See fig. 5-16.)

(b) The player propels the ball by kicking it with the feet or any part of the legs, by butting it with the head, and by hitting it with any portion of the body except the arms or hands.

(c) The goalkeeper is the only individual allowed to place hands on the ball, but only in the goalkeeper's area. The term "hands" includes the whole arm from the point of the shoulder down.

(d) A goal is made by causing the ball to cross completely the section of the goal line lying between the uprights and under the cross bar.

(e) Each goal scores 1 point for the team scoring the goal.

(f) The penalty for a foul committed anywhere on the playing field (except by the defensive team in its penalty area) is a free kick awarded to the opposing team.

(g) All opponents must be at least 10 yards from the ball when a free kick is taken.

(h) The penalty for a foul committed by the defensive team in its penalty area is a penalty kick.

(i) A penalty kick is a free kick at the goal from the spot 12 yards directly in front of the goal. The only players allowed within the penalty area at the time of the kick are the kicker and the defending goalkeeper.

(j) An official game consists of two 30-minute halves,

(k) Teams change goals at the end of every quarter.

(l) In the event of a tie, an extra quarter may played. This may be followed by a sudden death period. If the tie still exists, the game can be decided by penalty kick.

(m) After a team has propelled the ball across a side line and out of play, the ball is put back into play by a throw-in from the side line by a member of the opposing team. The ball is thrown in from the point where it crossed the side line as it went out of bounds. The ball must be thrown in with both hands using an over-the-head motion, keeping both feet on the ground.

(n) When the offensive team propels the ball across the defensive team's goal line but not in the goal, the defensive team is awarded a goal kick-a free kick taken from within its own goal area. The ball must come out of the penalty area to be in play.

(o) When the defensive team causes the ball to go behind its own goal line, excluding the portion between the goal posts, the offensive team is awarded a corner kick—a free kick taken by a member of the offensive team at the quarter circle, at the cornerflag post nearest to where the ball went behind the goal line. The flag post must not be removed.

(p) The game is started and, after a goal has been scored, is resumed by placing the ball in the center of the midfield line. Players must be on their side of the line until the ball is kicked. The ball must be kicked forward and must move at least 2 feet to be legal. The first kicker may not touch the ball twice in succession at the kick off. The opposing team must be 10 yards from the ball until it moves.

b.Speedball. Speedball offers vigorous and varied action with plenty of scoring opportunities. It is easy to learn and provides spontaneous fun. Little equipment is needed-a ball is all that is absolutely necessary. Speedball combines the kicking, trapping, and intercepting elements of soccer; the passing game of basketball; and the punting, drop kicking, and scoring pass of football.

(1) Place in the Program.Speedball, like soccer, should be introduced into the physical training program during the latter part of the toughening stage and used as a competitive activity in the sustaining stage. Speedball may be played any time weather permits but is primarily a spring or fall activity.

(2) Method of Play.The game is played as follows:

(a) Two teams of 11 each play the game under official rules, but any number of players may successfully constitute a team.An inflated leather ball, usually a soccer ball, is used. The playing field is a football field with a football goal post at each end. (See fig. 5-17.)

(b) The game starts with a soccer-type kickoff. The kicking team tries to retain possession of the ball and advance it toward the opposite goal by passing or kicking it. Running with the ball is not allowed, with the result that there is no tackling or interference. After the ball touches the ground, it cannot be picked up with the hands or caught on the bounce but must be played as in soccer until it is raised into the air directly from a kick. Then the hands may be used.

(c) When the ball goes out of bounds over the sidelines, it is given to a player of the opposing team and is put into play with a basketball throw- in. When it goes over the end line without a score, it is given to a player of the opposing team who may either pass or kick it onto the field.

(d) When two opposing players are contesting possession of a held ball, the official tosses the ball up between them as in basketball.

(e) Points are scored by kicking the ball under the crossbar of the goal post, drop kicking the ball over the crossbar, or completing a forward pass into the end zone for a touchdown.

Figure 5-17 is available in the illustration section at the end of the book.

(3) Basic Skills

(a) Shared Skills. Skills include kicking, passing, heading, and trapping (from soccer); punting, dropkicking and forward passing (from football); and passing, receiving, and pivoting (from basketball).

(b) Kickups. The kickup is a play in which a player lifts the ball into the air with the feet so as to legally play the ball with the hands. The kickup generally makes the transition from ground play to aerial play. The technique of making the play depends upon whether the ball is rolling or stationary. If the ball is rolling or bouncing toward the player, the foot is held on the ground with the toe down until the ball rolls onto the foot. Then the foot is raised, projecting the ball upward. If the ball is stationary, the player rolls it backward with one foot. Then the player places a foot where the ball will roll onto it. The player then raises the foot, projecting the ball as before. If a ball is rolling away, the player should stop it with a foot and play it as a stationary ball. The player can also raise the ball by standing over it with a foot on either side. The player presses the feet against the ball and jumps into the air, propelling the ball into the hands.

(4) Offensive and Defensive Play. The positions of players in speedball are much the same as in soccer. However, some of the positions have different names. Each team has 11 players. The forward line is composed of five players, the right end, right forward, center, left forward, and left end. The second line consists of right halfback, fullback, and left halfback. In the next line is the right guard and left guard. The player who defends the goal is the goal guard. The strategy employed in speedball during offensive play is very similar to that of soccer. There are two types of defensive formations in speedball: man-for-man and position defense. Man-for-man defense is recommended for beginning players.

(5) Abridged Rules

(a) Field. A speedball field is 100 yards long with additional 10-yard end zones on each end and 55 yards wide, the same as a regulation football field.

(b) Players. Eleven players are on a team. The goal guard has no special privileges.

(c) Time. Ten-minute quarters. Two minutes between quarters. Ten minutes between halves. Five minutes for extra overtime periods.(Begin the first overtime by a jump ball at center, same goals; change goals in the event of a second overtime period.)

(d) Winner of Toss. The winner of the toss has the choice of kicking, receiving, or defending a specific goal.

(e) Kickoff. A kick off starts the first and second half of a game and starts play after a score. The kick off is made from any place on the 50-yard line. The kicking team must be behind the ball when it is kicked. The receiving team must stay back of its restraining line (10 yards distance) until the ball is kicked. The ball must go toward the opposing team's goal before it can be played. Kick off out of bounds goes to the opponent at that spot. A kick off touched by the receiving team before going out of bounds, with no impetus added, still belongs to the receiving team. A kick off, in possession and control of the receiving team and then fumbled out of bounds, belongs to the kicking team at that spot.

(f) Starting Second and Fourth Quarters. The ball is given to the team that had possession at the end of the previous quarter, from out of bounds, as in basketball.

(g) Second Half. The first-half receiving team kicks off at the beginning of the second half.

(h) Fly and Ground Ball. The most characteristic playing rule of speedball is the difference between a fly ball (or aerial ball) and a ground ball. A player is not permitted to touch a ground ball with the hands and must play it as in soccer. A fly ball is one that has risen into the air directly from the foot of a player (for example, a punt, drop kick, place kick, or kickup).Such a ball may be caught with the hands provided the catch is made before the ball strikes the ground again.

(i) Kickup and Overhead Dribble. A kickup is a ball that is so kicked by a player that he can catch it himself. A bounce from the ground may not be touched with the hand because it has touched the ground since being kicked. This rule prohibits the ordinary basketball dribble, but one overhead dribble (throwing the ball into the air and advancing to catch it before it hits the ground) is permitted.

(j) Out of Bounds. If a team causes the ball to go out of bounds over the sidelines, a free throw-in (any style) is given to the opposing team. When the ball goes over the end line without scoring, it is given to the opponents who may pass or kick from out of bounds at that point.

(k) Tie Ball. In case two players are contesting the possession of a held ball, even in the end zone, a tie ball is declared. The ball is tossed up between them.

(6) Scoring

(a) Field Goal(3 points).A soccer-type kick, in which a ground ball is kicked under the crossbar and between the goal post from the field of play or end zone, is a field goal. (A punt going straight through is not a field goal for it is not a ground ball. The ball must hit the ground first.) A drop kick from the field of play that goes under a crossbar does not count as a field goal. A drop kick from the end zone that goes under the crossbar counts as a field goal;if it goes over the crossbar, it is ruled as a touch back, and the other team takes possession of the ball.

(b) Dropkick(2 points). A scoring drop kick must be made from the field of play and go over the crossbar and between the uprights. The ball must hit the ground before it is kicked (usually with the instep).

(c) End goal(1 point). This is a ground ball which receives its impetus (kicked or legally propelled by the body) from any player, offensive or defensive, in the end zone and which passes over the end line but not between the goal posts.

(d) Penalty Kick(1 point).A ball kicked from the penalty mark that goes between the goal posts and under the crossbar is a penalty kick. The penalty mark is placed directly in front of the goal at the center of the goal line.

(e) Touchdown(1 point). A touchdown is a forward pass from the field of play completed in the end zone. The player must be entirely in the end zone. If the player is on the goal line or has one foot in the field of play and the other in the end zone, the ball is declared out of bounds. If a forward pass is missed, the ball continues in play, but it must be returned to the field of play before another forward pass or drop kick may be made.

(7) Substitutions. Substitutions may be made any time when the ball is not in play. A player who is withdrawn may not return during that same period.

(8) Timeout. Three legal timeouts of 2 minutes each are permitted each team during the game.

(9) Fouls

(a) Personal Foul. Kicking, tripping, charging, pushing, holding, blocking, and unnecessary roughness of any kind, such as running into an opponent from behind, are personal fouls. Kicking at a fly ball and thereby kicking an opponent is a personal foul. Four personal fouls disqualify a player from the game.

(b) Technical Foul. Illegal substitution, more than three timeouts in a game, unsportsmanlike conduct, and unnecessarily delaying the game are technical fouls.

(c) Violation. Traveling with the ball, touching a ground ball with the hands or arms, double overhead dribble, violating tie ball, and kicking or kneeing a fly ball before catching it are violations.

(d) Penalties. The offended player shall attempt the kick.

(e) Officiation of Fouls

1 Fouls in the field of play allow no follow-up while fouls in the end zone always allow follow-up.

2 On penalty kicks, with no follow-up, only the kicker and goalie are involved.

3 On penalty kicks, with a follow-up, the kicking side is behind the ball and the defending side behind the end line or in the field of play. No one is allowed in the end zone or between the goal posts except the goal guard. The kicker must make an actual attempt at goal and cannot play the ball again until after another player plays it.

c. Volleyball. Volleyball is a popular sport. The game entails much physical activity, yet it is not strenuous. It is, therefore, a game for young and old alike, for beginners and for skilled players. It may be played indoors or outdoors on any type of terrain. As an informal activity, volleyball can be played by any number; as an organized activity, it provides, as few other sports do, a game for 12 to play in a limited area. While volleyball requires no great skill to play, it permits a high degree of proficiency. Players naturally get more enjoyment when they know the game and play it well. For this reason, instruction in the basic skills should be provided.

(1) Place in the Program Volleyball may be used occasionally as a competitive activity during the sustaining stage. Though it is a year- round sport, it should be included in the physical training program only when a more strenuous activity is impractical. Volleyball is an excellent off-duty activity.

(2) Instruction. Usually during the first three or four classes, a 10- to 15- minute period of instruction, followed by scrimmage, is enough to teach the basic skills, rules, and techniques of volleyball. More time, if available, can be given to teaching basic skills, but emphasize competitive play rather than formal instruction. Divide the unit into 6-person teams, organized on the basis of ability .Teams should be as nearly equal as possible.

(3) Basic Skills

(a) Passing the High Ball. The chest pass is the most effective method of playing the ball. To receive the ball, the knees are fixed, and the body is tilted forward. The elbows are raised sideward to a point in line with the shoulders. The wrists are extended in line with the forearm. The arms, wrists, and hands are rotated inward. To pass the ball, the hands are chest-high, thumbs pointing inward. The fingers are flexed, forming a cup, allowing them to contact the ball. On contact with the ball, the wrists are snapped while the fingers and elbows are pushed upward, sending the ball upward. A high ball is much easier to handle than a low one.

(b) Passing the Low Ball. A ball that is lower than the waist is one of the easiest to hit, but it is also a frequent cause of fouls (holding or carrying the ball). The best position for handling a low ball is to have the knees flexed and arms flexed at the elbows and rotated so the thumbs are pointing outward, the palms up. When the fingers contact the ball, the entire body reacts in a lifting motion. The arms and hands swing upward in a scooping action. It is important that the fingers, not the palms, contact the ball, and that the ball is batted, not thrown.

(c) Underhand Serve. Take a position behind the back line facing the net, left foot forward, holding the ball in the palm of the left hand. The left knee is flexed; the right knee is straight. Swing the right arm back and at the same time move the left hand (holding the ball) across the body in line with the right hip. Then swing the right arm forward hitting the ball off of the left hand with the palm of the right hand, raising the hips and arching the back in the same motion. Be certain to swing the right arm in a straight line, or the ball will be difficult to control. When the opposition is in formation, the server should try to place the ball in the right or left back area, and not near the net.

(d) Setup. A setup is a ball hit into the air near the net by one player, so that a teammate may hit or "spike" it sharply downward into the opponent's court. The chest pass is the best pass to use. The ball is sent approximately 10 feet into the air toward the spiker so it will descend from 4 to 20 inches from the net.

(e) Spiking. The spike is a leap into the air and a sharp downward hitting of the ball into the opponent's court. A spiker must be able to spring easily from the floor, judge the movement of the ball, and strike it with a downward movement of the arm. To jump from the floor, step off with one foot and jump with the other. Stand with the right or left side to the net, facing the setup player. Much depends upon the setup player to place the ball in the proper position. The spiker jumps into the air and strikes the ball above its center so as to drive it downward. A snapping movement of the arm and wrist will drive the ball forward and downward with power and control. Aim for a weak spot in the opponent's defense.

(f) Blocking. The block is a defense technique used to prevent a spiker from driving the ball across the net. It is an attempt by one or more defensive players at the net to block a hard-hit shot by using the force of the ball to send it immediately back into the opponent's court. In an effective block, forwards on the defensive team spring into the air at the time of the spike, placing both hands and arms in the expected path of the ball. An effective block tends to upset the offense and presents another element for the spiker to worry about. To be effective, the blocker must anticipate the path of the ball and time the block with the spike.

(4) Abridged Rules

(a) The volleyball court is 30 feet wide by 60 feet long. (See fig. 5-18.)

(b) The top of the net is 8 feet high.

(c) A volleyball team consists of 6 players.

(d) A match consists of the best two out of three games.

(e) The first team scoring 15 points wins the game, provided that they have 2 points more than their opponents.

(f) A deuce game is a game in which both teams score 14 points. The game is continued until one team obtains a 2-point advantage over the other.

(g) Only the serving team can score. If the serving team commits a fault, it loses the serve to the opposing team.

(h) The team members who receive the served ball rotate one position in a clockwise direction.

(i) The ball is put into play by serving from behind the back line.

(j) A served ball touching the net results in the loss of the serve. At any other time during play, a ball touching the net is still in play.

Figure 5-18 is available in the illustration section at the end of the book.

(k) The ball is out of play when it touches the ground or goes outside one of the boundary lines.

(l) All line balls are good.

(m) The players must hit or bat the ball; they may not throw, lift, or grasp it.

(n) A player may not touch the ball with any part of the body below the knees.

(o) A player may not play the ball twice in succession. In receiving a hard-driven spike, a defensive player may make several contacts with the ball even if they are not simultaneous. All such contacts, however, must constitute one continuous play, and all must be above the knees.

(p) The ball may be touched no more than three times on one side of the net before being returned across the net to the opposing team.

(q) A player must not touch or reach across the net.

(r) A player may touch the line under the net, but may not cross it.

Chapter 6
EVALUATION OF PERFORMANCE DURING TRAINING

6001.OBJECTIVES OF COMBAT READINESS

The objectives of the Marine Corps physical fitness program are-

a. To contribute to the health and well-being of every Marine through regular exercise and health education.

b. To develop Marines who are physically capable of performing their duties in garrison and in combat.

c. To develop Marines with a reserve level of physical fitness that will help them win in combat.

d. To provide a medium for developing the self-confidence of the individual Marine and thereby enhancing overall discipline, morale, esprit de corps, unit efficiency, and the desire to excel within the Marine Corps.

6002. THE PHYSICAL FITNESS TEST

The physical fitness test (PFT) has distracted attention from the proper goal of the physical fitness program. Simply stated, the goal of the physical fitness program is the success of Marines in combat. Unit commanders must not allow the PFT to become an end in itself. The PFT should be used within the commander's program of combat readiness training as a standard, an index of the physical fitness of individual Marines at a given time and place. The PFT is not difficult, and it must be kept in perspective. All athletes should be first class, and first class is attainable for all Marines. The conduct of the PFT is described in Marine Corps Order 6100.3_.

6003. OTHER METHODS OF EVALUATION

The commander should use other ways of evaluation to provide variety and imagination. Intangible objectives, such as confidence and aggressiveness, are hard to measure. The use of either inspection or observation is necessary to evaluate objectives of this nature. The commander has several other methods available when evaluating the physical condition of the command. The additional methods of inspection, observation, medical examination, and testing are summarized as follows:

a. Formal Inspections. Formal inspections, using inspection officers and standardized rating criteria, may help evaluate unit physical fitness.

b. Daily Observation. Routine observation of physical performance and appearance can serve as an indicator of an individual's or unit's physical readiness. However, mere observation is not a totally reliable or accurate means of evaluation.

c. **Physical Examination.** Medical examination may detect any individual disability or detrimental physical condition. It may guide in application of remedial, therapeutic, or limited exercise programs.

d. Other Physical Fitness Tests. Some other tests are described in detail in this manual for information purposes. Commanders are cautioned, however, not to place too great an emphasis on testing.

6004. RESPONSIBILITY

The commander is responsible for the physical fitness of the command and for the measurement and evaluation of its physical readiness. The goal of combat readiness training remains to prepare Marines and units for the physical demands of combat. The bulk of the commander's combat readiness training program should be such activities as obstacle and confidence courses, endurance courses, and progressive load-bearing marches which will prepare Marines for combat. Group activities which are competitive and combative in nature are advantageous and add imagination and enthusiasm to the program. Activities such as orienteering, which is not only physically demanding but also competitive and which teaches essential subject skills, are excellent and motivate all concerned. A commander's goal should be to provide an integrated training program that is so aggressive and demanding of Marines and leaders alike that a PFT administered at any time will reflect that each and every Marine is both physically and mentally prepared for the demands of combat.

6005. COMMANDER'S PHYSICAL FITNESS PROGRAM

The Commandant of the Marine Corps has directed that the physical fitness test shall be the universal measure of adequate individual physical fitness. The focus of a commander's physical fitness training program, however, should not be to prepare Marines to pass the physical fitness test, but to prepare them for the physical rigors of combat.

a. Administration. The physical fitness program requires each Marine to spend a minimum of 3 hours per week in physical fitness training, to be tested semiannually, and to obtain a minimum standard of third class. The attainment of a higher score is a laudable individual goal which should be encouraged but should not become a unit objective. Overemphasis could be detrimental to training required to develop the complete Marine. All commanders should exercise sound judgment as well as positive, aggressive leadership in striving toward the training objective.

b. Variety of Activities. The program must include the following elements:
(1) Physical Readiness Training. physical readiness training is a complete physical training program which develops and maintains the strength, endurance, and physical skills needed to sustain the individual during combat.
(2) Remedial Physical Conditioning. Remedial physical conditioning is a process by which physically substandard Marines are conditioned to meet prescribed standards.
(3) Competitive Conditioning Activities. Competitive conditioning activities consist of teams or individuals competing against an opponent to win. This involves a combination of sports and military skills designed to foster competitive attitudes and develop unit pride and esprit de corps.
(4) Occupational Conditioning. Marching long distances, dry net training, etc., should not be ignored as a means to fulfill the time requirement.

c.Time. In order to develop the desired level of physical fitness, unit commanders are advised that a minimum program of 3 hours per week is mandatory. The program must encompass at least three exercise periods weekly. Each period should include calisthenics, running, or other forms of vigorous activity .The minimum desirable length for each exercise period is 60 minutes. Shorter periods may be authorized by the unit commander if dictated by the local training situation. Unit commanders are encouraged to use the normal working day to satisfy this training requirement. However, commanders are authorized to conduct required physical fitness training during off-duty hours when the mission, workload, personnel status or other significant factors preclude fulfilling it during duty hours.

d.Action. Marine Corps Order 6100.3_ requires that commanders-

o Establish and maintain an effective physical fitness program which ensures that all Marines maintain an acceptable level of physical fitness.

o Ensure that all individuals are medically qualified prior to participating in the physical fitness program.

o Establish a minimum physical fitness program of at least 3 hours per week. Physical training may be authorized on an individual basis at the discretion of the unit commander.

o Ensure that all Marines participate in physical fitness conditioning activities commensurate with their medical qualifications and limitations.

o Conduct physical fitness testing for all Marines.

o Place all Marines who fail to pass the PFT on a daily command-supervised remedial physical conditioning program until they pass the test.

o Ensure that results of physical fitness testing are entered on section A of the fitness report. Comments are placed in section C in connection with attainment of superior physical performance or a medical excuse from the PFT.

6006. OTHER FORMAL TESTS

Commanders must realize that the best index of combat readiness is personal observation of performance on conditioning marches of 15 miles while under load. This manual presents three additional indicators of unit fitness. These indices are the physical readiness test, which was formerly the Marine Corps standard and is now routinely administered at Officer Candidate School; the airborne trainee physical fitness test which many Marines who attend U.S. Army formal schools are required to take; and the battle fitness test, the ultimate test of the combat readiness of Marines. Commanders may wish to include these tests or portions thereof within their combat readiness programs.

6007. PHYSICAL READINESS TEST

a. Purpose. The purpose of this test is to measure the Marine's ability to meet the minimum standard of physical readiness for duty in the field. This test is currently administered to all male officer candidates at Officer Candidate School, MCCDC, Quantico, VA. Female candidates do not take this test.

b. Test Area. Administration of this test requires an area suitable for crawling, jumping, and running 160 yards to include a trench that is 8 feet wide.

c. Equipment. The equipment required for this test includes—
o A platform, rail on a ledge, or a box 18 inches high,
o A 20-foot climbing rope,
o A stopwatch.

d. Conduct of Test. Adequate timers and other supervisory personnel must be provided by the company staff. All events should be conducted in a single session of one morning or afternoon. The events may be run in any sequence.

e. Uniform. The uniform for the physical readiness test is as follows: full utilities, cartridge belt, belt suspender straps, two canteens (full of water), rifle, and helmet. Depending on the heat condition, the PT shirt may be worn in place of the utility jacket. All equipment will remain secure throughout the test. Except for emergencies, canteens are not to be emptied.

f. Events

(1) Event 1: Climbing Uphill (Step-ups)This event simulates marching uphill at a rapid and steady pace. The Marine stands in front of the step. On the command GO, the Marine places either foot on the step and steps up. Hand pressure on the knee may be used if desired. The Marine then stands erect and steps backward and down, on one foot at a time. The same leg may be used for each repetition or the legs may be alternated. For 100 points, Marines will complete 60 repetitions in the maximum allotted time of 1 minute and 20 seconds (80 seconds).

*(2) Event 2: Rope Climb. T*his event resembles entering and leaving a hovering helicopter, using ropes in house-to-house fighting, and ascending and descending landing nets. The Marine assumes a sitting position on the ground at the bottom of the rope with hands at the highest point that can be reached. On the command GO, the Marine jumps up and climbs to the top. The Marine must touch the 20-foot mark with one hand, then descend without sliding or dropping. Marines must reach the 20-foot mark in the maximum allotted time of 30 seconds or less.

(3) Event 3: Evacuation. This event simulates reaching and evacuating a wounded Marine under fire. The Marine assumes the prone position at a distance of 50 yards from a casualty who is of approximately equal weight. On the command GO, the Marine springs up and covers the distance in a sprint. The Marine lifts the casualty to a fireman's carry and returns to the starting point. Assistance by the casualty is permitted. The tested individual must carry all equipment; the casualty has none. Marines must negotiate the course in a maximum time of 47 seconds or less.

(4) Event 4.'Advance by Fire and Maneuven. This event simulates advancing and assaulting during an attack. The Marine assumes the prone position at the starting point. On the command GO, the Marine then alternately crawls, runs in a zigzag manner, rolls, jumps, and assumes different firing positions for the next 130yards. Before reaching the objective, the Marine must "hit the deck," roll over, and assume the firing position three times. While assuming the firing position, the Marine must pull back the charging handle, take aim, and squeeze the trigger before rising from the ground. After rising from the ground for the third time,the Marine continues running and jumps to clear the 8-foot trench in one leap. The Marine then finishes the course in

the maximum allotted time of 2 minutes. The Marine is allowed multiple attempts only within the specified time limit.

(5) Event 5: Forced March. This event is a combat run of 3 miles without halts and against time. The Marine must double-time. Marines must complete the 3-mile course in a maximum allotted time of 30 minutes or less.

g. Event Standards

(1) To successfully pass the physical readiness test, the Marine must pass each event with a minimum of 80 points for each. Points are then assessed for each event in accordance with the scoring tables. (See fig. 6-1.)

(2) If a Marine fails one event, the highest possible score that the Marine may receive for the entire test is 79 percent. (See example 1 in fig. 6-2.) If two events are failed, the highest possible score for the entire test is 69 percent, regardless of the total points accumulated. (See example 2 in fig. 6-2.) If three events are failed, the highest possible score is 59 percent. Failure in four or more events will result in a score of zero for the entire test.

(3) To convert the total points for all five satisfactorily passed events to a final percentage score, total the accumulated points for all five events and divide by five. (See example 3 in fig. 6-2.)

Figure 6-1 is available in the illustration section at the end of the book.

Figure 6-2 is available in the illustration section at the end of the book.

6008. AIRBORNE TRAINEE PHYSICAL FITNESS TEST

a. Use and Composition of the Test.

(1) Use. The airborne trainee physical fitness test is a means of determining the physical ability of the applicant for acceptance to and retention in the airborne course of instruction.

(2) Test Events The test battery consists of five events as follows: chinups, knee bender, pushups, situps, and an endurance run.

b. Method of Scoring and Standards

(1) Scoring. The Marine will be scored by a trained scorer who is thoroughly familiar with the minimum standards for the test events. The Marine will be scored on a pass or fail basis. The performance on each event may be recorded on the reverse side of DA Form 705 (Physical Fitness Testing Record).The examinee records the personal information on the face of the card as specified on the first, second, and third lines. The face of the card is identified by writing diagonally across the lower half of the card "Airborne Trainee PPT—See reverse side."Using line 7 on the reverse side of the scorecard, the scorer enters the test title in the block entitled "(Other (Specify))" and completes the required information on the remainder of that line. The "Remarks" section can be used to record the test event titles and the applicant's performance on each test event.

(2) Standards. To successfully pass the test, the examinee must reach the standard in each test event. The standards follow:

(a) Chinups

(b) Knee bender 80(2-minute period)

(c) Pushups 45(2-minute period)

(d) Situps45(2-minute period)

(e) Endurance runs:2 miles in 15:54 minutes or less in athletic gear and 4 miles in 32 minutes in utilities and running shoes.

c. Uniform for Testing

(1) Examinees. The prescribed uniform for test participation is boots and the work uniform of the season. No headdress is worn. When climatic conditions permit, jackets or outer shirts may be removed.

(2) Officials. Scorers and other test officials should be uniformly and distinctively dressed for contrast with Marines being tested.

d. Test Administration Procedure

(1) Preparation for Test. The administration of the test to a large group makes it mandatory that the test effort be organized and efficiently operated. All testing is not completed with large test groups; on certain occasions, individuals and small groups are tested. Care must be exercised to administer the test uniformly and to standardize the conduct of all elements of the test. Regardless of the size of the test group, the following elements of sound test administration should be included:

(a) An orientation to include the purpose, method of administration, scoring of the test, preparation of the score card, and required standards.

(b) A correct demonstration of each event to ensure that there is no misunderstanding of the proper form and required standards.

(c) Completion of all five test events in one test period with all Marines taking the events in the same sequence.

(d) Adequate rest periods between the test events to allow for recovery before the next event.

(2) Method of Administration with a Large Group. With a minimum of 12 lanes per test event, 14 officials can administer the test battery to 150 or 200 Marines in 2 hours. The officials are designated as follows: one officer in charge, one demonstrator,and 12 scorers. If more or fewer Marines are to be tested, a greater or lesser number of officials will be required. The following procedure is recommended:

(a) Conduct an orientation and ensure the examinees have properly completed their score cards.

(b) Assign Marines to lanes and caution them to remain in the same lane order throughout the test.

(c) Explain and demonstrate the chinup event, administer it, and score it. Then proceed to the knee bender and pushup events, and administer them in the same manner.

(d) Grant a 5- to 10-minute rest period after the pushup event. Advise against excessive consumption of water during the break period.

(e) Explain and demonstrate the situp event, administer it, and score it as prescribed. Then move to the run area, explain the running event, and complete it.

(f) Retain the scorecards at the completion of the running event.

(3) Method of Administration with a Small Group. A similar procedure is followed for the testing of individuals and small groups. The informality usually associated with small groups must not conflict with sound test administration. With fewer examinees, fewer officials are required.

---

e. Description and Explanation of Test Events

(1) Test Event l:Chinups

(a) Purpose. This event is devised to test arm and shoulder flexor strength.

(b)Equipment There is one horizontal bar per lane, made of plumber's pipe or a gymnasium horizontal bar 1 1/2 inches in outside diameter. The bar must be rigidly supported at a height of 8 feet above the ground. The upright supports must be 5 feet apart. There must be a movable stand at each bar for short Marines to stand on to reach the bar.

(c) Officials There is one scorer per lane.

(d) Organization. The examinees, holding their scorecards, stand in order behind the restraining line in their respective lanes. The scorers take each scorecard when the Marine is called forward for the test.

(e) Starting Position The bar is grasped with the palms turned toward the face, the thumbs underneath the bar. The body is fully extended in a "dead" hanging position with the arms straight and the feet above the ground.

(f) Movement. Pull the body directly upward until the chin is placed over the bar. Lower the body until the elbows are completely straight and the body is again in the "dead" hanging position. Repeat as many times as required.

(g) Instructions. Explain and demonstrate the fully extended "dead" hanging position with the proper grasp. Show that the chin is placed over the bar at the top of the movement and that the arms are fully extended, the elbows completely straight, at the bottom of the movement (the hanging position). Explain that the body must be kept from swinging and that it is permissible to raise the legs and flex the hips when pulling up, but that any kicking, bicycling, or jerking motion with the trunk or legs is not acceptable. Inform the examinees that no penalty is exacted for hanging on the bar to rest in the bottom position but that this is not to their advantage. Tell them that half-completed chinups are not counted, and that the scorer will repeat the number of the last correct chinup when incorrect execution is detected.

(h) Administration and Scoring. Caution the examinees to assume the "dead" hanging position and wait for the scorer's command to begin. The scorer is at the examinee's left with a clear view of the bar. If the examinee begins to swing widely, the scorer should stop the swinging by extending the left arm across the front of the examinee's body, being sure not to hinder the execution of the chinup. The scorer counts aloud the number of chinups correctly executed. When a chinup is not correctly executed, the scorer repeats the number of the last correct one. The scorer records the number of correct chinups on the scorecard and returns the card to the examinee.

(2) Test Event 2:Knee Bender

(a) Purpose. This event measures the strength and endurance of the leg muscles.

(b) Equipment. None.

(c) Officials There is one scorer per lane.

(d) Organization. Marines stand in numerical order behind the restraining line in their respective lanes. The scorer takes each scorecard when the Marine is called forward for the test.

(e) Starting Position. The feet are spread less than shoulder width apart, hands on hips, thumbs in the small of the back, elbows back.

(f) Movement. Do a knee bend and at the same time bend slightly forward at the waist and thrust the arms between the legs until the extended fingers touch the ground. The hands are about 6 inches apart. The bend is approximately a three-quarters bend. From this knee bend position, recover to the starting position by moving the body upward, straightening the knees, and returning the hands to the hips. Repeat as many times as required.

(g) Instructions. Explain and demonstrate the correct starting position. Be certain examinees understand the correct knee bend and that only the tips of the fingers touch the ground. Tell them the scorer will repeat the number of the last correct knee bender when incorrect execution is detected. Some of the common errors are failure to correctly bend the knees, failure to touch the ground, and failure to assume the completely erect position after the bend has been executed.

(h) Administration and Scoring. The scorer stands to one side so as to see that the knees are properly bent and the fingers touch the ground as prescribed. From this position, the scorer can view the examinee to see that a properly erect position is assumed after each knee bend. The scorer counts aloud the number of correctly executed knee bends. When a knee bend is done incorrectly, the scorer repeats the number of the last correct one. The scorer records the number of correct knee bends on the scorecard and returns it to the examinee.

(3) Test Event 3:Pushups

(a) Purpose. Pushups measure arm and shoulder extensor strength.

(b) Equipment. None.

(c) Official. There is one scorer per lane.

(d) Organization. Marines stand behind the restraining line in their respective lanes until the scorer calls on them to perform. The scorer takes each scorecard when the Marine comes forward.

(e) Starting Position. The front-leaning rest position is the starting position. The body is straight from head to heels, palms are flat on the ground directly underneath the shoulders, and elbows are straight and locked. The body weight is supported on the hands and toes throughout the event.

(f) Movement. Bending only the elbows, lower the body in one straight plane until the chest touches the scorer's hand. Straightening and locking the elbows, raise the body in one straight plane, returning to the original front-leaning rest position. Repeat as many times as required, keeping the body in a straight line from head to heels.

(g) Instructions. Explain and demonstrate that the arms are straight at the beginning and completion of the movement and that the chest must touch the scorer's hand, but the stomach and thighs must not touch the ground. Also explain that the whole body must be maintained in a straight line as it is lowered and raised; that is, there is to be no breaking at the hips or shoulders so that any body part is lowered or raised in advance of the other or as a separate segment .Dipping or rolling through the shoulders is illegal, as is lowering or raising the body with one arm or shoulder at any time. Resting is not permitted during repetitions. Instruct

Marines that the scorer will repeat the number of the last correct pushup when incorrect execution is detected.

(h) Administration and Scoring. It is recommended that Marines assume a prone position while placing their feet and hands in the proper positions. This permits them to rest while the scorer gets into position and, at the same time, provides a feel of the body in a straight plane from head to heels. The scorer lies on the right hip and side to the right of the examinee. This gives the scorer a clear view of the examinee's body to see any errors. The palm of the scorer's right hand rests flat on the ground underneath the lowest part of the examinee's chest. By keeping the right forearm flat on the ground at an angle in front of the examinee's right arm, the scorer's position will not prevent the examinee from lowering the body completely. The scorer's left hand is free to test the straightening of the elbow at the completion of the movements and to point out body segments being lowered or raised separately. When in position and ready, the scorer has the examinee assume the starting position and begin doing pushups. The scorer counts aloud the repetitions done correctly and repeats the number of the last correct pushup if one is done incorrectly. There is no penalty if the contour of the examinee's body causes the hips to protrude slightly out of line, provided that the whole body is raised and lowered simultaneously. The scorer enters the number of repetitions on the scorecard and returns it to the examinee.

(4) Test Event 4: Situps

(a) Purpose. This event primarily measures abdominal strength.

(b) Equipment. None.

(c) Official. There is one scorer per lane.

(d) Organization Marines stand behind the restraining line in their respective lanes until the scorer calls on them to perform. The scorer takes each scorecard when the Marine comes forward.

(e) Starting Position. The examinee lies flat on the back with knees flexed, both feet flat on the ground. The correct angle of the thighs to the ground is 45 degrees. If the heels are too near the buttocks, the applicant will not be able to sit up. Legs are spread shoulder width apart. The examinee interlaces fingers and places them behind the head in contact with the ground. The feet are not held by another person.

(f) Movement. Bend forward at the waist and raise the upper body until the head is directly over the knees. Heels are not to leave the ground. Elbows remain in the same plane to the head and body throughout the event. The upper body is slowly lowered to the starting position until the head touches the ground. Repetitions are done at a slow cadence with no rest periods.

(g) Instructions. Explain and demonstrate the correct starting position and the proper execution of the situps to be sure that Marines understand the movement. Warn them that their knees must remain flexed during each situp, the heels cannot leave the ground at any time, and they may not roll up on one side and push up with one elbow. Tell them they must do the repetitions at a slow cadence with no rest periods. Instruct Marines that the scorer will repeat the number of the last correct situp when incorrect execution is detected.

(h) Administration and Scoring. When the performer is in position and ready, the scorer has the performer assume the starting position and begin doing situps. The scorer counts aloud the correct executions. When a situp is improperly done, the scorer repeats the number of the last correct one. No situp is credited if the hands are unclasped from behind the head, if the back is used to bounce up from the ground (which means the shoulders would not touch the ground), or if one shoulder or elbow is used to push up. The examinee is not penalized if the heels slide forward slightly so long as the knees remain flexed and the heels maintain contact with the ground. The scorer enters the number of repetitions on the scorecard and returns it to the examinee.

(5) Test Event 5:Endurance Run

(a) Purpose. This event measures cardiovascular endurance.

(b) Equipment One stopwatch or watch with a sweep second hand.

(c) Area. A large training field on which a quarter-mile track has been staked out or a level road over flat terrain may be used as a running surface. A 1-mile route is designated with wooden stakes marking the start point, finish point, and one-quarter mile intervals.

(d) Officials For large groups, a scorer times the event and controls the conduct of the run, and a guide runs with the group and sets the pace.

(e) Organization The run is conducted with groups of Marines in a column formation. Company-sized units may run at the same time with platoons serving as running groups. The scorer issues the command to assume double-time.

(f) Starting Position. Marines are assembled in proper column formation (column of twos, threes, or fours, as appropriate to the size of the group) with short men to the front. When all is ready, the column is moved forward a short distance before the running period is started.

(g) Movement. At the command DOUBLE-TIME, MARCH, the examinees retain their places in the column formation and execute the command. Length of steps is about 40 inches. The scorer has the group execute the run. The formation is maintained during therun.

(h) Instructions. Marines are instructed to maintain formation while running and are informed that the guide will set the proper pace. They will be instructed in the command used to control the column in the execution of the test. The scorer should announce the 4-minute, 2-minute, 1-minute, and 30-second remaining time intervals.

(i) Administration and Scoring. The event may be administered as previously prescribed to a large group, to several Marines, or to an individual. An individual examinee usually does not require a guide or pacer. If the event is administered on a training field, the scorer may stand in the center as the runner(s) circles about the field. This method of administration relieves the scorer of running with each group to be tested. Scoring is based on successful completion of the run as prescribed. The scorer should announce the remaining times as prescribed in (h).

6009. BATTLE FITNESS TEST

The battle fitness test is a physical fitness evaluation which is currently under development as part of the basic warrior training concept plan.

Chapter 7
THE HUMAN BODY

7001.GENERAL

To implement an effective physical conditioning program, a leader must understand how the human body functions and how exercise affects the body .This chapter provides only an introductory discussion of anatomy and functioning.

a. Body Functioning During the Stages of Conditioning. For more on the stages of conditioning, see chapter 1.

(1) During the toughening stage, the waste products of muscle activity (lactic acids) collect more rapidly than the blood can remove them. This acid waste builds up in the muscle tissue and irritates the nerves in the muscle fiber, causing pain and stiffness. As the exercise program continues, more blood is carried through the muscle, removing the waste materials more rapidly and eventually causing the soreness to disappear.

(2) During the slow improvement stage, the blood circulation in the muscles increases, and the body as a whole becomes a more efficient machine. The improvement is rapid in the first few weeks, but as a higher level of skill and conditioning is reached, the improvement becomes less noticeable. The body reaches its maximum level of performance after 6 to 10 weeks and should then be maintained at this peak.

(3) During the sustaining stage, it is possible to maintain this state of conditioning through 15 to 20 minutes of exercise a day, but the exercise must be quite strenuous.

b. Diet. Regular exercise increases the appetite. If the desire for greater amounts of food is satisfied by a balanced diet, the body benefits.

(1) There are two main types of foods: body-building and energy- producing.

(a) Body-building foods consist of proteins, which build up tissue and repair wear and tear.

(b) Energy-producing foods include carbohydrates and fats. Carbohydrates provide a quick source of energy, while fats act as a reserve store of energy.

(2) In addition, food contains vitamins, mineral salts, and water. During hot weather and strenuous training periods, the body requires greater fluid intake.

(3) Diet should be supplemented with proper rest to allow the digestive system to digest the lood.

(4) Occasionally, especially during early stages of conditioning, strenuous exercise may cause vomiting. Although vomiting is not a frequent occurrence, it is not usually a cause for concern.

7002. SYSTEMS OF THE BODY

The systems of the body include the skeletal, muscular, circulatory, respiratory, endocrine, digestive, genitourinary, and nervous systems. Each has a different function, but all must work in cooperation with one another to insure a sound body. Of these systems, the first four are the most affected by exercise and are discussed in succeeding paragraphs.

7003. THE SKELETON

a. Bones. The skeleton (see fig. 7-1) is composed of about 206 bones, which-
o Provide structure for the body and support for the attachment of muscles,
o Protect vital organs, such as the brain, lungs, and heart,
o Manufacture red blood cells, which carry oxygen through the body.

Figure 7-1 is available in the illustration section at the end of the book.

b. Joints. A joint is a place of union between two or more bones. Joints can be-
o Immovable, such as in the face and head.
o Slightly movable, such as in the vertebrae or sternum.
o Freely movable, such as in the shoulder, hip, knee, ankle, and elbow. The bones in a freely movable joint are held in place by ligaments. Freely movable joints are of the greatest importance in physical training because exercise improves their mobility and stability.

c. Cartilage and Ligaments. The joints in the body, where bones connect, are supported by cartilage and ligaments.
o Cartilage is a tough, elastic, translucent tissue that acts as a shock absorber or buffer between bones. Examples are the discs between the vertebrae, the tissue attaching the ribs to the sternum, and buffers in the knee joints. Damaged cartilage does not heal.
o Ligaments are connective tissue that bind bones together. They contract and extend but are not elastic. Consequently, ligaments can be torn or strained. Damaged ligaments will heal to some extent.

7004. THE MUSCLES

a. General. Muscles are tissue, or an organ composed of tissue, which contract or extend to produce movement. Muscles are of three classifications: involuntary, voluntary, and cardiac. The voluntary and cardiac muscles are the most affected by physical training; physical training has little effect on involuntary muscles.(Seefigs. 7-2, 7-3, 7-4.)
o People have no control over the involuntary muscles, an example of which is the diaphragm,
o Voluntary muscles are the larger skeletal muscles which cause movement and which the individual can control.
o Cardiac muscle is found only in the heart and for all practical purposes is an involuntary muscle. Physical training has a significant effect on the fitness of cardiac muscles.

b. Muscle Structure. Muscle is composed of cells which are grouped into bundles called fibers. Groups of fibers form muscles. These bundles of fibers are held in place by sarcolemma, a thin, sheath-like material that surrounds the muscle bundles and secretes a fluid that lubricates the muscle tissue. The fused ends of the sarcolemma form the tendons which attach the muscles to the bones.

c. Attachment of Muscles. The arrangement of muscles on the skeleton provides the proper angle of pull to make movement possible. Voluntary muscles are usually attached to the skeleton in two places.

d.Action of Muscles. To produce motion and to do work a muscle usually shortens its fibers. The movement may be flexion such as bending the arm at the elbow, or extension, such as the straightening the arm. Some muscles raise the arms or legs, others depress the raised limb. Some muscles have the primary function of rotating the trunk from side to side. In addition, muscles act as stabilizers as well as prime movers.

Figure 7-2 is available in the illustration section at the end of the book.

Figure 7-3 is available in the illustration section at the end of the book.

Figure 7-4 is available in the illustration section at the end of the book.

7005. FUNCTIONING OF THE SKELETON AND MUSCLES

a. General. The skeleton and muscles function in close coordination. Muscles move the body, and bones support the muscles.

b. Effect of Exercise on Bones. Continuous exercise, particularly among younger people, usually strengthens the bones, allowing them to withstand greater stress and strain. Bones which are not used lose much of their strength, a factor which should be considered when individuals return to physical training after a prolonged period of inactivity. People in this category should refrain from activities which might result in bone breakage before the bone is exercised back to normal condition. The condition known as "march fracture" is usually a result of this condition.

c. Muscular Strength. When exercised vigorously to improve strength, muscle grows in size. In general, the larger the muscle, the stronger the muscle. Furthermore, trained muscle functions more efficiently than untrained muscle. Trained muscles contract more vigorously and with less effort. Regular and strenuous exercise also toughens muscle, making it firmer and able to stand more strain.

d. **Muscular Endurance.** Muscles naturally become fatigued during continuous, repeated work. Through specific exercises, such as chinups or situps, local muscle groups become exhausted long before an individual fatigues. Training to develop muscular endurance enables people to continue a relatively heavy load of exercise over a long period of time. Lengthened exercise periods improve muscular endurance.

e. Muscular Coordination. Improved speed and strength result in part from improved muscular coordination. An unskilled individual may use irrelevant muscles to perform a particular activity, increasing the amount of work without increasing the mechanical output. This increase in skill is highly desirable, but it should be offset by greater effort in duration or intensity to compensate for the loss in overload due to increased skill.

f. Muscular Fatigue. When the rate of work is increased, the energy required is proportionately much greater than the increase in rate. For example, if an individual doubles running speed, the amount of power demanded to do this is increased eight times.

g. Circulation in Muscles Regular, strenuous exercise causes the creation of new capillaries and the opening of inactive, latent capillaries, increasing blood circulation within the muscle by as much as 400 percent. This increases the supply of food materials and oxygen to the muscle, improving its endurance. This process takes about 8 to 12 weeks of regular conditioning in young adults and longer as age increases.

7006. THE CIRCULATORY AND RESPIRATORY SYSTEMS

a. The Circulatory System. The circulatory system transports blood to all parts of the body, removes waste products for disposal, and delivers protecting and repairing substances where needed. The circulatory system includes the heart and blood vessels.

(1) The Heart. The heart is a pump which forces blood through the blood vessels to the parts of the body. The heart is a little larger than a fist and is located in the left center of the chest between the lungs.

(2) Blood Vessels. The vessels carrying blood away from the heart are the arteries. They eventually divide into capillaries, the very small vessels through which diffusion and osmosis take place. The capillaries gradually increase in size, forming the veins which carry blood back to the heart.

b. The Respiratory System. The respiratory system performs the function of breathing, whereby oxygen is inhaled and carbon dioxide is exhaled. The respiratory system consists of the mouth, nose, trachea, lungs, and diaphragm.

o Trachea. The trachea, or windpipe, is a hollow, tube-like structure that carries air from the mouth to the lungs.

o Lungs. The lungs are elastic bags in the chest where the exchange of oxygen and carbon dioxide takes place.

o Diaphragm. The diaphragm is a thin, sheet-like muscle just below the lungs. During inhalation, the diaphragm flattens out and lowers, allowing the lungs to expand and fill with air. During exhalation, the diaphragm raises into a dome shape, helping to expel air from the lungs.

7007. CARDIOVASCULAR FUNCTIONING

a. General. Cardiovascular functioning is the combined functioning of the circulatory and respiratory systems. The chief organs of these systems, the heart and lungs, function together to provide oxygen to the body.(See fig. 7-5.)

Figure 7-5 is available in the illustration section at the end of the book.

b. Heart Action. The heart, the organ which propels blood through the blood vessels, is the chief organ of cardiovascular endurance. A person tires quickly if the heart lacks the capacity to circulate the blood through the body. The heart is a muscular organ and is developed by exercise just like any other muscle. The heart cannot be exercised alone; any exercise that puts a sufficient load on the heart, such as running or cycling, will also exercise other body parts. A conditioned heart tends to beat more slowly but pump more blood when at rest. This is known as an increase in "stroke volume,"a desirable condition because it enables the heart to pump more blood with a slower contraction rate.

c. Functions of the Lungs. Strenuous and regular exercise can improve the efficiency with which the lungs transmit oxygen to the blood by as much as 25 percent In a poorly conditioned person, some of the alveoli (air sacs) within the lungs are closed or collapsed. The forced breathing created by exercise over a period of weeks causes the air sacs to slowly expand, increasing the ability to absorb oxygen.

d. Relationship of Heart and Lungs. The heart pumps carbon dioxide-laden blood through the pulmonary artery to the lungs. In the lungs the carbon dioxide is exchanged for oxygen and the purified blood is returned to the heart by way of the pulmonary vein. The heart then pumps the blood through the aorta for circulation throughout the body. Blood moving into the muscles exchanges oxygen for carbon dioxide, after which time it is ready for the return trip to the heart.

e. Cardiovascular Functioning in High Altitudes. Marines to be employed in high altitudes should be acclimated in a similar area for 10 to 14 days prior to employment. Persons not accustomed to the rarefied air of higher altitudes tire more quickly and may collapse after rapid physical exertion. Air is much less dense at high altitudes than at sea level. Persons accustomed to sea level inhale only about 80 percent of the oxygen they are used to no matter how hard they breathe. Further, people accustomed to sea level or moderate altitude simply do not have enough red corpuscles in the blood to fulfill their needs at high altitudes. The red blood cells will increase over time.

f. Symptoms of Cardiovascular Malfunctioning. The brain is the first organ to react to a lack of oxygen. Unconsciousness results when the brain is denied sufficient oxygen. "Blacking out" is actually a defense mechanism to keep the body alive.

APPENDIX A
INSTRUCTOR HINTS AND INSTRUCTOR TRAINING

1. General. This appendix informs instructors on administration of exercise periods, commands, the extended rectangular and circle formations, methods of instruction and conduct of exercise activities, explanation of basic positions associated with the various activities, and instructor training.

2. Aim of the Physical Fitness Program. The aim of the physical fitness program is to prepare Marines physically for war. This can be conveniently split into two phases.

a. Preparation Phase

(1) Development of all-around physical fitness.

(2) Development of purposeful physical skills.

(3) Development of mental alertness.

(4) Development of character and leadership.

b. Maintenance Phase. Once the preparation is completed, the program must progress to a maintenance program. This phase revolves around the need to maintain the trained Marine's physical fitness for a particular role. This will include—

(1) Maintenance of a high standard of all-around physical fitness.

(2) Application of purposeful physical skills to combat skills.

(3) Development of leadership and self-confidence.

(4) Development of mental toughness and endurance.

3. Physical and Mental Development

a. Marines need to be developed physically and mentally. The physical training (PT) instructor must ensure that the program accomplishes this.

b. The system of the body functions best in response to progressive exercise. This means that Marines should master the easier exercises before they attempt the more difficult ones.

c. The body and mind are linked inseparably and must be considered as one in physical fitness training. Physical exercises assist in developing the mind.

d. The confidence in oneself and in one's comrades attained by physical achievement helps to promote the indomitable will to win. The Marine must be taught to realize this. It will help during arduous training, which at all times must be imaginative, realistic, and filled with enthusiasm. All physical fitness training must be associated with developing the qualities required of a Marine.

4. Planning and Preparation of Programs The following factors must be considered when compiling a program (see also ch. 2)~

a. Aim or desired skill level.

b. Duration of the complete training program and number of physical training lessons.

c. Purpose of included exercises,drills, or practical applications.

d. Beginning/inventory fitness levels.

e. Medical facilities, medical personnel, and Marine Corps or local command regulations governing their proximity to the training.
f. Number of Marines to be trained.
g. Availability of instructors, assistants, and facilities.
h. Apparatus and equipment available.
i. Alternative training areas for varying weather conditions,
j. Stages of progression and tests of achievement.

5. Five Basic Principles of Program Design
a. Regularity. Regularity of exercise is far more important than the amount of exercise performed on a given day. A Marine should exercise a minimum of three times per week. Human muscle begins to atrophy after 3 days without exercise.
b. Balance. The program must be balanced in terms of total physical fitness. Imbalance is created when, for example, an individual merely runs. While the cardiovascular system is improved, general upper body strength may be lacking. Conversely, a person who just does weight training may develop strength, but not cardiovascular endurance.
c. Overload. If total body fitness is to be developed, then the muscular and cardiovascular systems must be made to perform more work than previously. This is done by increasing the frequency, intensity, and duration of the exercise. The body adapts to the amount of stress placed upon it and becomes stronger and more efficient.
d. Progression. To be successful, any type of training program must be progressive. This is an area where untrained persons conducting PT can cause injury to Marines resulting in failure of the PT program. Close supervision must be made of untrained instructors. The result of progressing too fast is that the class begins to reject PT because it is no longer enjoyable or motivating. The program should start at a slow pace, and the pace should increase as proficiency improves. Average persons will encounter three stages of progressiom the toughening stage, the slow improvement stage, and the sustaining stage. (See ch. 1.)
e. Variety. Some programs fail because the training becomes boring. Variety is essential and perhaps presents the greatest challenge to the commander and staff.

6. Grouping. Providing for different levels of physical fitness is particularly recommended in the early stages of conditioning. Older Marines and those in poorer physical condition should be expected to attain a group level of fitness, but they should be given more time to do it.
a. Homogeneous Grouping One simple method of providing for the difference in levels is to group them according to their condition. A two-group classification would divide Marines into highly conditioned and average groups. A finer classification could be obtained by dividing them into three groups—a highly conditioned, a moderately conditioned, and an unconditioned group.
b. Performance or Age Grouping. The segregation of Marines into different exercise groups should be based on physical fitness test scores or on the level of fitness they demonstrate. They may also be grouped at first according to age. A common classification by ages is under 30, 30 to 34, and 35 and above.

7. Preparatory Commands and Commands of Execution. The prepreparatory command describes and specifies what is required. The command of execution calls into action what has been prescribed. All preparatory commands are given with a rising inflection. The interval between commands is long enough to permit the average Marine to understand the first one before the second one is given.

8. Extended Rectangular Formation. The traditional formation for carrying on many physical training activities is the extended rectangular formation (fig. A-1). This formation is more compact than any other. It is the best type to use for large numbers of Marines because it is simple and easy to assume.

a. For the formation of oneplatoon, the base man paces off five paces from the stand, faces left and moves five paces, halts, and again faces left. With the base man positioned facing the stand, the platoon leader then commands. **FALL OUT AND FALL IN ON THE BASE MAN**. At this command, all Marines run to the designated area and reform. This procedure is preferred to marching the unit into position. If more control is desired, the unit may march at double time to the vicinity of the base man and then be directed to fall out and fall in on him. Much valuable time is wasted in the field by needless maneuvering of Marines at quick time in an effort to position the platoon or unit on the exact spot for the exercises.

b. A company-size unit assumes the extended rectangular formation from a column of threes or fours at normal intervals between squads. This extension can also be executed from a company mass without interval between platoons. In extending either a platoon or company-size unit, take your place at the head of the column and command:

(1) TAKE INTERVAL TO THE LEFT, MARCH. At this command, the Marines in the right flank file stand fast with arms extended sideward. All others turn to the left and run forward at double time. After taking a sufficient number of steps, all Marines face the front with both arms extended sideward. The distance between finger-tips is about 12 inches, and dress is right.

Figure A-1 is available in the illustration section at the end of the book.

(2) ARMS DOWNWARD, MOVE. At this command, the arms are lowered smartly to the sides.

(3) LEFT, FACE

(4) EXTEND TO THE LEFT, MARCH. At this command, those in the right flank file stand fast with arms extended sideward. All others turn to the left and run forward at double time. Spacing is the same as in (1) above, and dress is right.

(5) ARMS DOWNWARD, MOVE. Same as in (2) above.

(6) RIGHT, FACE.

(7) FROM FRONT TO REAR, COUNT OFF. At this command, the leading Marine in each column turns the head to the right rear, calls of **ONE** and faces the front. Successive Marines in each column call off in turn **TWO, THREE, FOUR, FIVE** in the same manner.

(8) EVEN NUMBERS TO THE LEFT, UNCOVER. At this command, each even-numbered Marine stride jumps sideward to the left, squarely in the center of the interval. In doing this, each one swings the left leg sideward, jumps from the right foot to the left foot, and smartly brings the right into position against the left.

c. To assemble the unit, you commandASSEMBLE TO THE RIGHT, MARCHAt this command, all return to their original position in the column at double time and reform on the base man.

d.It is recommended that the area for grounding equipment and arms be at the edge of, or well away from, the area to be used for exercising. To conserve time and ensure proper position of the unit, the base man or, if the unit is composed of several platoon-sizegroups, the various base men may precede the unit and establish their positions in relation to the instructor's stand.

9. Circle Formation.The circle formation is effective for conducting various exercise activities (fig. A-2). This formation has an advantage in that supervision of all Marines is facilitated and a moving formation is available which provides control. Guerrilla exercises, grass drills, and some forms of running are examples of activities which are more easily conducted in the circle formation than in the extended rectangular formation.

a. It is not advisable to have more than 60 Marines in a circle. When more must be accommodated, separate circles should be used.Concentric circles may be employed to reduce the size of the circle or to accommodate more persons. If concentric circles are formed, the different circles are made by designating squads for each circle.Each additional circle requires more Marines than the one inside it.For example, one squad of a platoon may form the inner circle and theremaimng three squads the outer circle.

b. When a platoon is to form a circle, the commands ai€IRCLEFORMATION, MARCH, FOLLOW MEUpon this command, the right flank squad of the column moves forward at double time with the leader of the platoon group gradually forming a circle in a counterclockwise direction. Each succeeding file falls in behind that on the right. After the rough outline of the circle is formed, the leader commandsPICK UP A **5**-YARD INTERVAL.

Figure A-2 is available in the illustration section at the end of the book.

c.The group may be halted and faced toward the center, or, if instruction is not necessary, the exercise activity may be executed without stopping the platoon.

lO.Leadership Techniques

a. Unless you experience all the exercises, you cannot appreciate how arduous they are, what movements are most strenuous and difficult, where the errors in performance are likely to occur, and what the proper cadence should be.

b. You must give everyone careful supervision and participate in the exercises to show that you can do them. When you participate, your assistant instructors should supervise because it is difficult for you to supervise and exercise at the same time.

c. Marines should never be kept too long in one position, especially a constrained one. They should never have to perform an exercise more times than they can do it without losing the proper form. You should insist upon proper form in the execution of all exercise activities.Even slight deviations from the proper form reduce the value of the exercise.

d. Avoid long explanations.As a rule, it should be necessary to give a full explanation of new exercises only. Explain the most essential features of an exercise first; add details later. Too many details at one time are more likely to confuse than to assist them.Minor corrections should be made to the entire class while the exercise is in progress (for exampHEADS **UP,** KNEES STRAIGHT). If necessary, follow this direction by the name of the person who is particularly at fault. If a Marine requires special attention, givethat person separate instruction by an assistant instructor to avoidwasting the time of the group.

e. The heavy demand on your voice can be lightened by training assistant instructors to assume some of the instruction. Using mass cadence is also an effective method of lessening the demand on your voice.

“Illustrations”

DAY	TIME	ACTIVITY
MONDAY	60 MIN	RUNNING ACTIVITY
TUESDAY	60 MIN	COMBATIVES
WEDNESDAY	60 MIN	CONDITIONING & LOG DRILLS
THURSDAY	60 MIN	CONDITIONING & RELAYS
FRIDAY	60 MIN	RUNNING & CONDITIONING
SATURDAY	60 MIN	RUNNING ACTIVITY

Figure 2-1. Sample A- -Single Activity.

DAY	TIME	ACTIVITY	REMARKS
MONDAY	20 MIN 20 MIN 20 MIN	RIFLE DRILL COMBATIVES RUNNING	PROGRESSIVE ACTIVITY SYSTEM
TUESDAY	30 MIN	GUERRILLA EXERCISES TABLE I	SINGLE ACTIVITY
WEDNESDAY	30 MIN	RIFLE DRILL	SINGLE ACTIVITY
THURSDAY	30 MIN	CIRCUIT INTERVAL TRAINING	SINGLE ACTIVITY
FRIDAY	20 MIN 20 MIN 20 MIN	STRENGTH CIRCUIT CONDITIONING DRILL CROSS-COUNTRY RUN	ROTATION ACTIVITY
SATURDAY	60 MIN	WARM UP RUN AND GRASS DRILL	SINGLE ACTIVITY

Figure 2-2. Sample B- -Various Time Periods, Activities, and Systems.

DAY	TIME	ACTIVITY
MONDAY	45 MIN	RUNNING – CROSS-COUNTRY
TUESDAY	75 MIN	TEAM ATHLETICS – TOUCH FOOTBALL
WEDNESDAY	45 MIN	RUNNING – CROSS-COUNTRY
THURSDAY	20 MIN 40 MIN	LOG DRILL DUAL COMBATIVES
FRIDAY	90 MIN	SPEED MARCH
SATURDAY	90 MIN	TEAM CONTESTS

Figure 2-3. Sample C--Sustaining Activities.

Route	Kilometers from SP	Recommended rate of march (kmph)	Remarks
SP: RJ 8th Div Rd—Superhighway	......	4	Bridge; hard surface; two guides.
RJ Jamestown—Hersey Rd	2.25	4	Traffic heavy; two guides; Jamestown Rd bears to right.
RJ Jamestown—Yankee Rd	7.10	4	Two guides
RJ Jamestown—Lightning Rd	10.50	4	Two guides
RJ Jamestown—Sunshine Rd	12.40	4	Light traffic; one guide Sunshine Rd; gravel, poor traction when wet.
RP; RJ Sunshine—Sedan Rd	14.80	4	Two guides

Figure 3-1.Example of Route Reconnaissance Report.

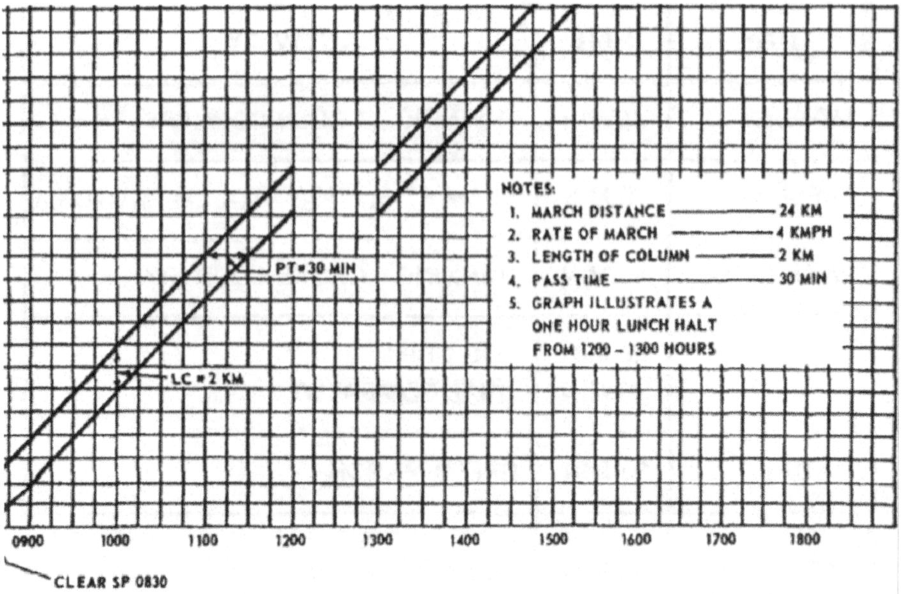

Figure 3-2.Road Movement Graph.

Figure 3-3.The Pace-Setter.

	SINGLE FILE	COLUMN OF TWO
2 m/MAN APART	24	12
5 m/MAN APART	54	27

GAP DISTANCES			
SERIAL:	(3-1) × 100 m	=	200 m
MARCH UNIT:	(12-1-2-) × 50 m	=	450 m
	TOTAL COLUMN GAP		650 m

Figure 3-4.Conditioning Drill One.

Figure 3-5.Conditioning Drill Two.

Figure 3-6.Conditioning Drill Three.

Figure 3-7. Rifle Drill.

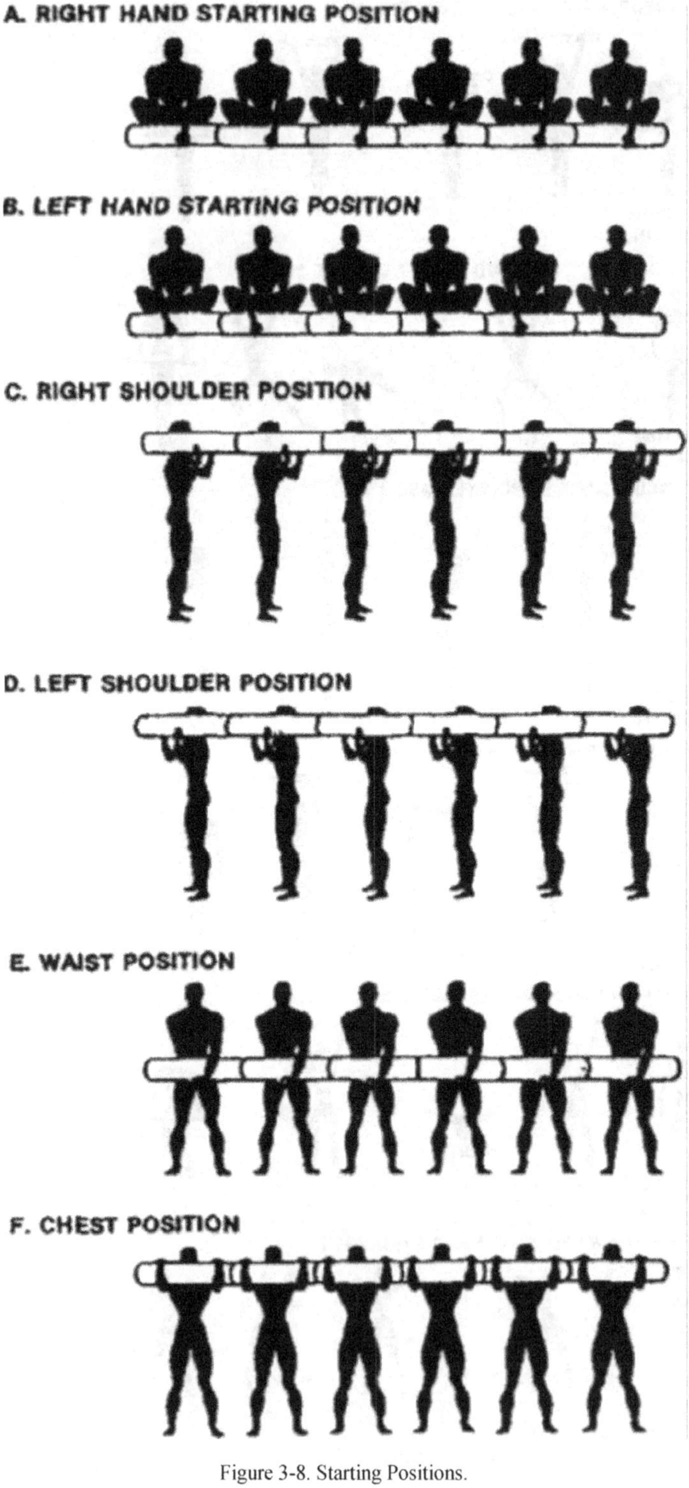

Figure 3-8. Starting Positions.

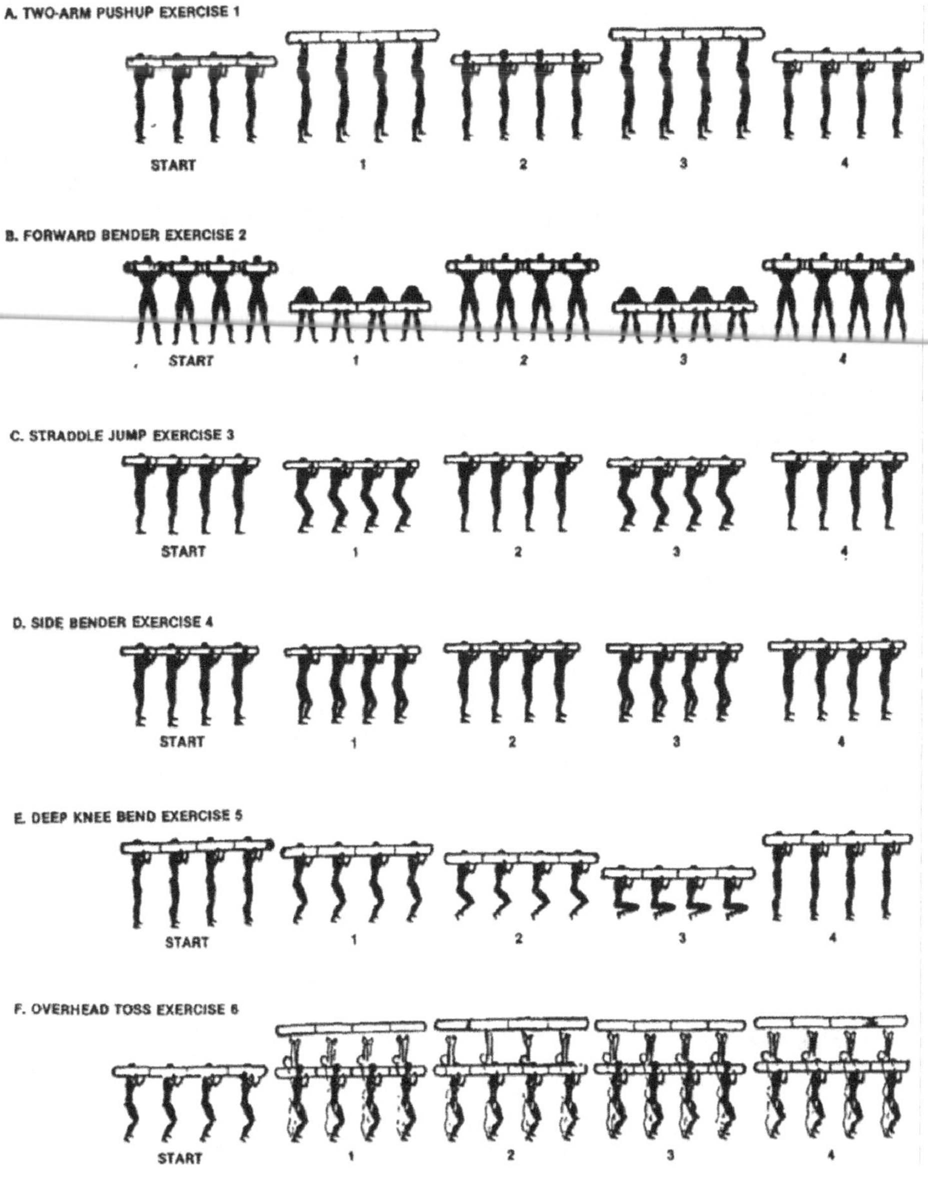

Figure 3-9. Log Drill

Figure 3-10.Basic Positions for Grass Drill.

Figure 3-11.Grass Drill One.

Figure 3-12.Grass Drill Two.

Figure 3-13.Guerrilla Table I.

Figure 3-14.Guerrilla Table II.

Figure 3-15.Proper Running Form.

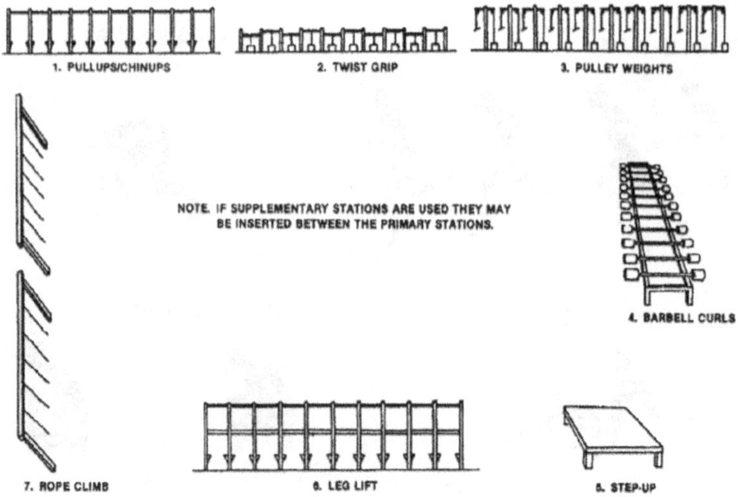

Figure 3-16. Fixed Strength Circuit.

STATION	ITEM	NO.	SPECIFICATION
1	BARBELL		1¼-INCH PIPE 5 FEET LONG WITH CONCRETE FILLED NO. 10 CANS
2	JUMP ROPE		¼- OR ⅜-INCH ROPE, 10 FEET LONG
3	TWIST GRIP		HANDLE 12 INCHES LONG, ROPE 4 FEET LONG, NO. 10 CAN CONCRETE FILLED.
4	INCLINE PLANE		¾-INCH PLYWOOD PLATFORM 2 FEET WIDE AND 6 FEET, 6 INCHES LONG, ELEVATED 10 INCHES AT ONE END. STRAP TO HOLD FEET DOWN.
5	WAR CLUB		HEAD IS 6 BY 12 INCHES, HANDLE IS 14 INCHES LONG BY 1¼ INCHES IN DIAMETER, ABOUT 20 POUNDS.
6	BICYCLE RIDE		PLYWOOD BOARD OR PLATFORM 2 BY 3 FEET WITH 2 BY 2 RUNNERS.
7	STEP—UP		A BOX OR STURDY PLATFORM 18 INCHES HIGH, 18 INCHES WIDE, 24 INCHES LONG.
8	ISOMETRIC PULL		TWO HANDLES 12 INCHES LONG WITH 4 FEET (BETWEEN HANDLES) OR LIGHT WIRE CABLE OR ¼-INCH ROPE

Figure 3-17.Movable Strength Circuit.

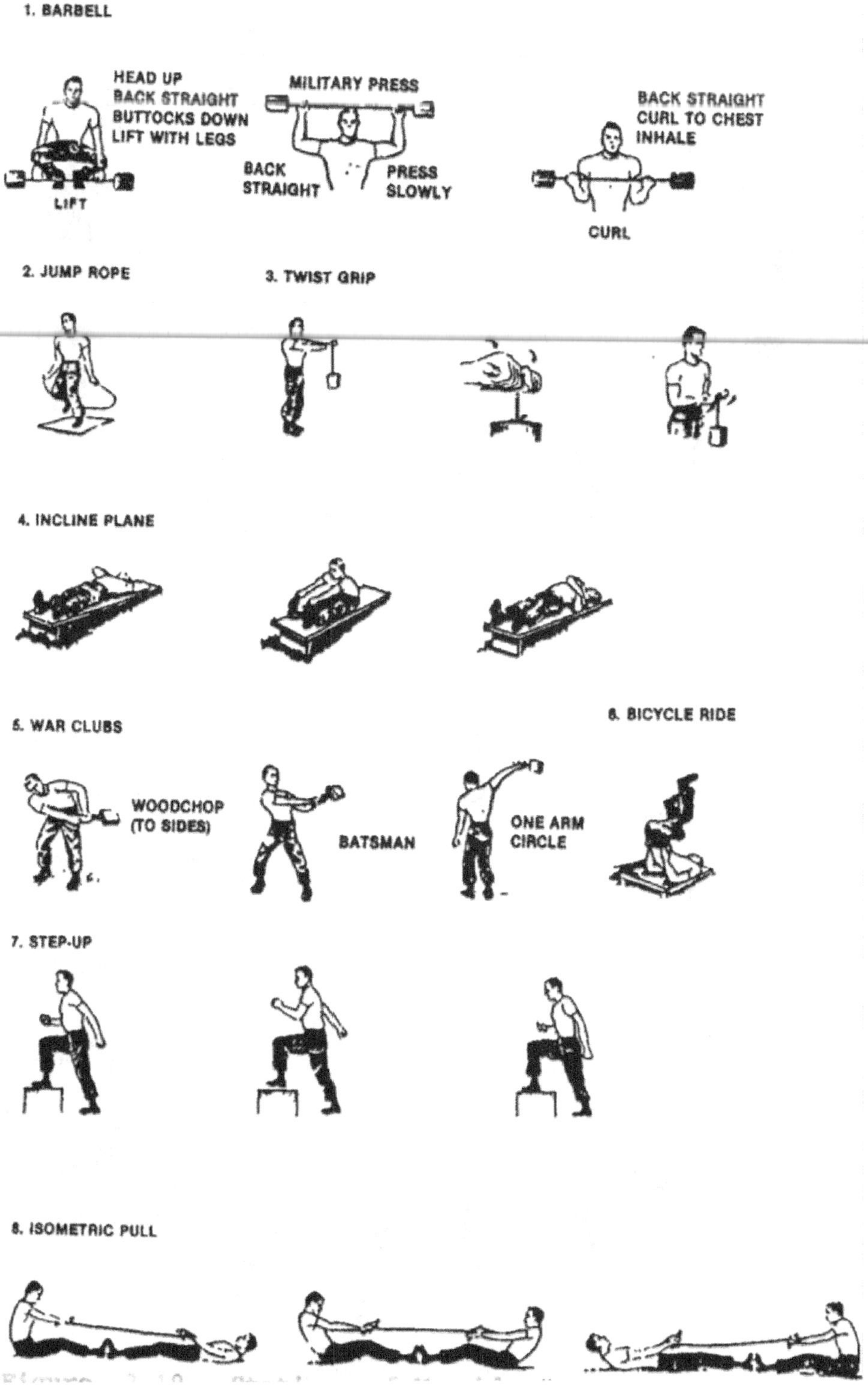

Figure. 3-18.Stations of Movable Strength Circuit.

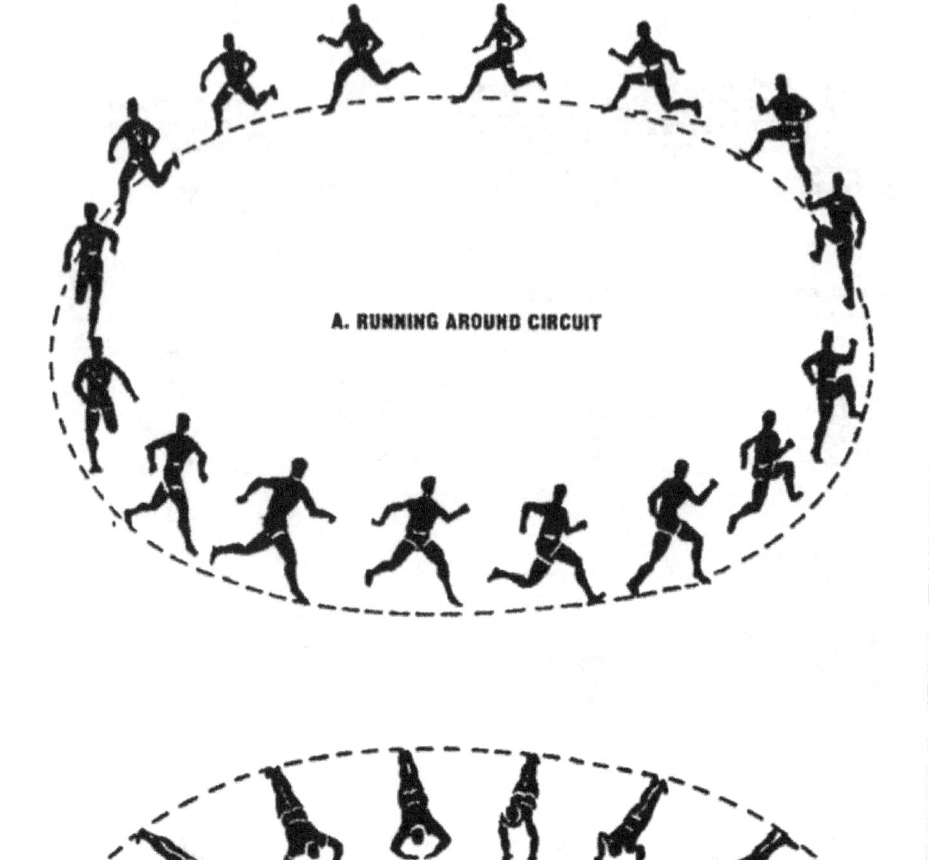

Figure 3-19. Circuit-IntervalTable.

A. BROAD JUMPING B JUMPING UPWARD C. JUMPING DOWNWARD

Figure 3-20. Jumping.

A. DODGE TO AVOID OBSTACLE B. CHANGE OF DIRECTION BEHIND CONCEALMENT C. DODGE TO AVOID DANGER AREAS

Figure 3-21.Dodging.

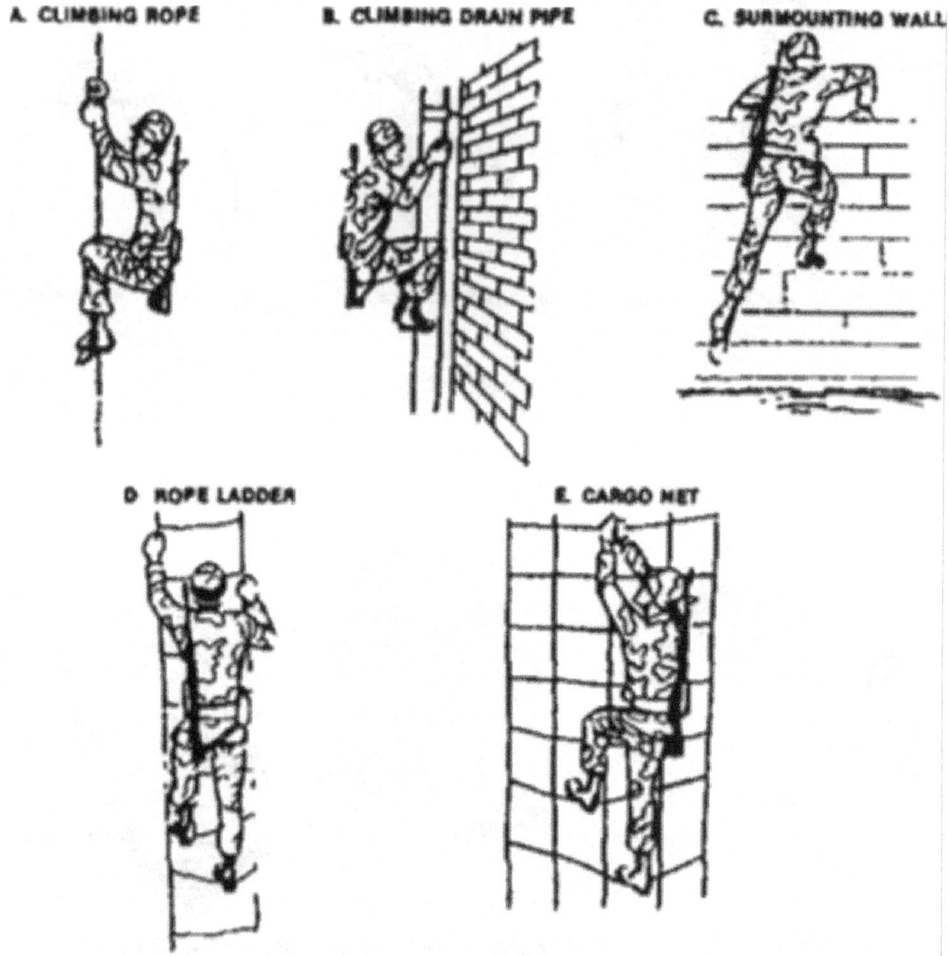

Figure 3-22.Climbing and Surmounting Vertical Objects.

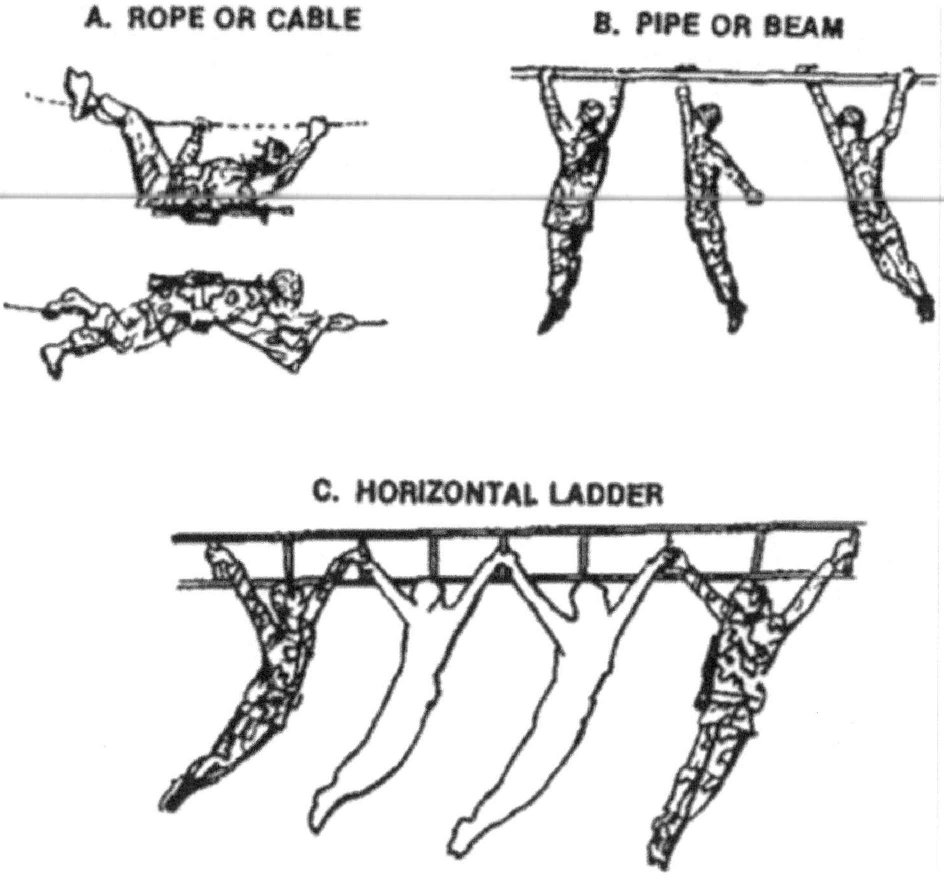

Figure 3-23. Traversing Horizontal Objects.

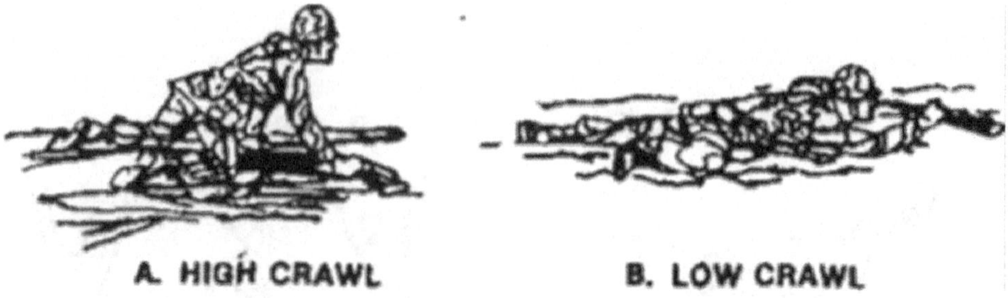

Figure 3-24.Crawling.

Figure 3-25.Throwing.

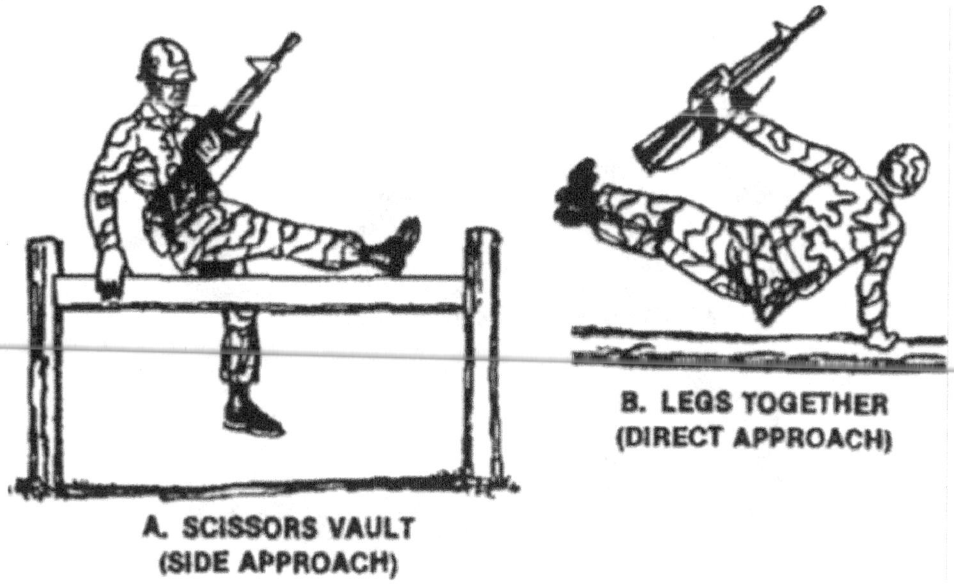

Figure 3-26. Vaulting

Figure 3-27.Balancing.

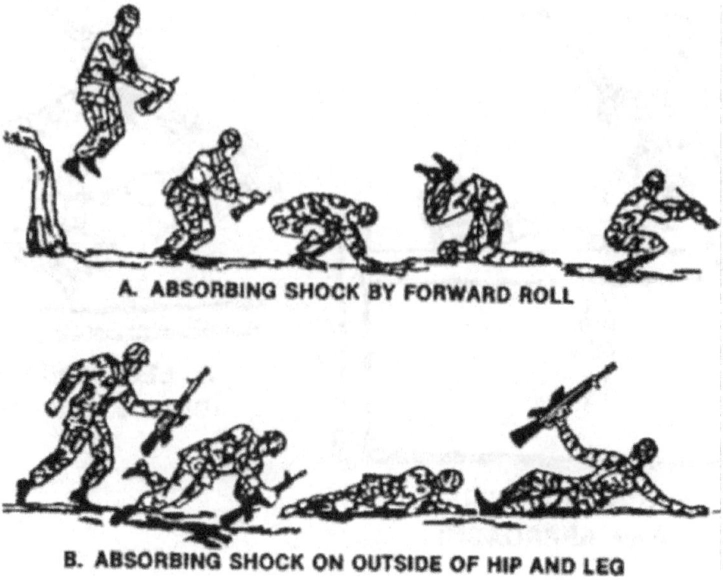

Figure 3-28.Falling.

Figure 3-29.Jumping-Type Obstacles.

LANES TO GUIDE CHANGE OF DIRECTION

MAZES TO CAUSE CHANGE OF DIRECTION

Figure 3-30.Dodging-Type Obstacles.

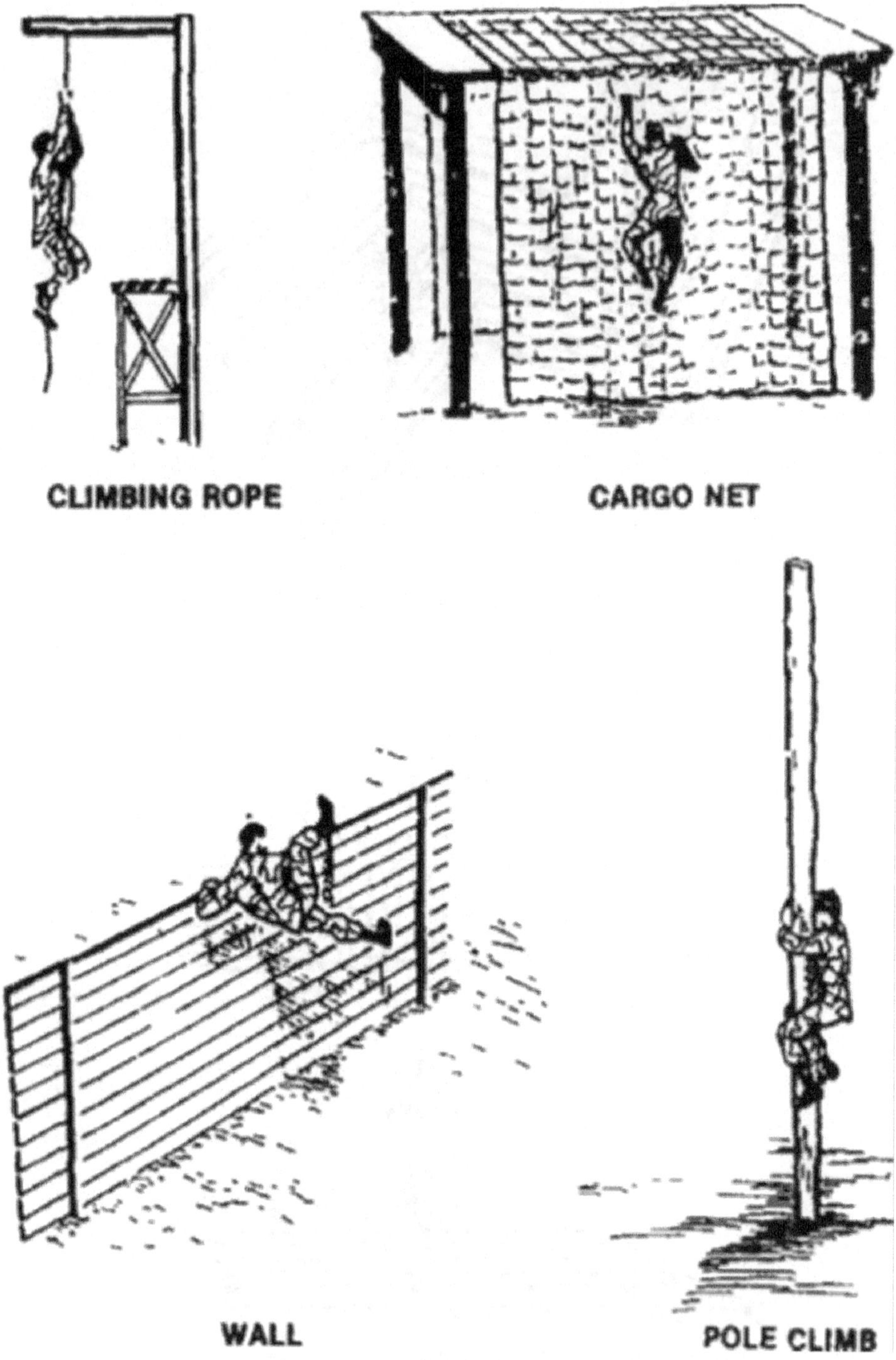

Figure 3-31. Vertical Climbing Obstacles.

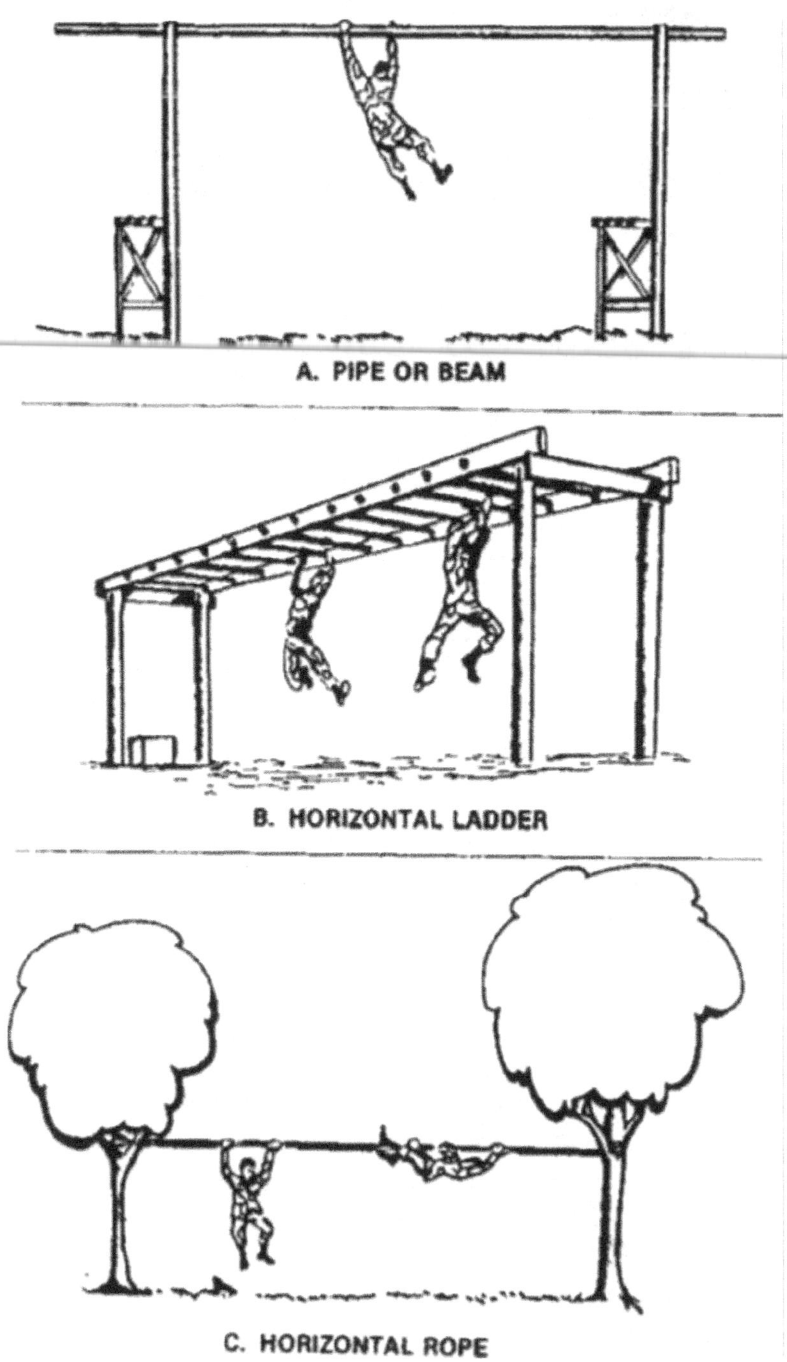

Figure 3-32.Horizontal Type Obstacles.

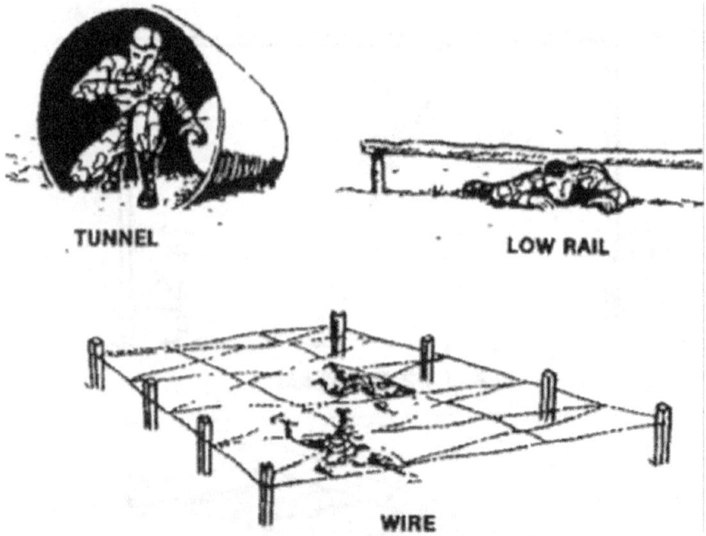

Figure 3-33.Crawling-Type Obstacles.

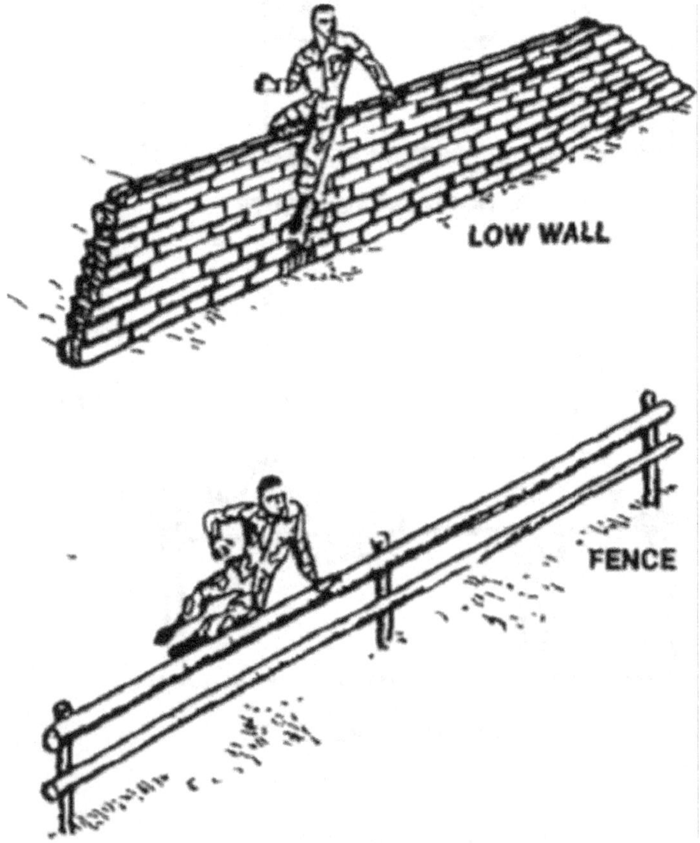

Figure 3-34. Vaulting Obstacles.

Figure 3-35.Balance-Type Obstacles.

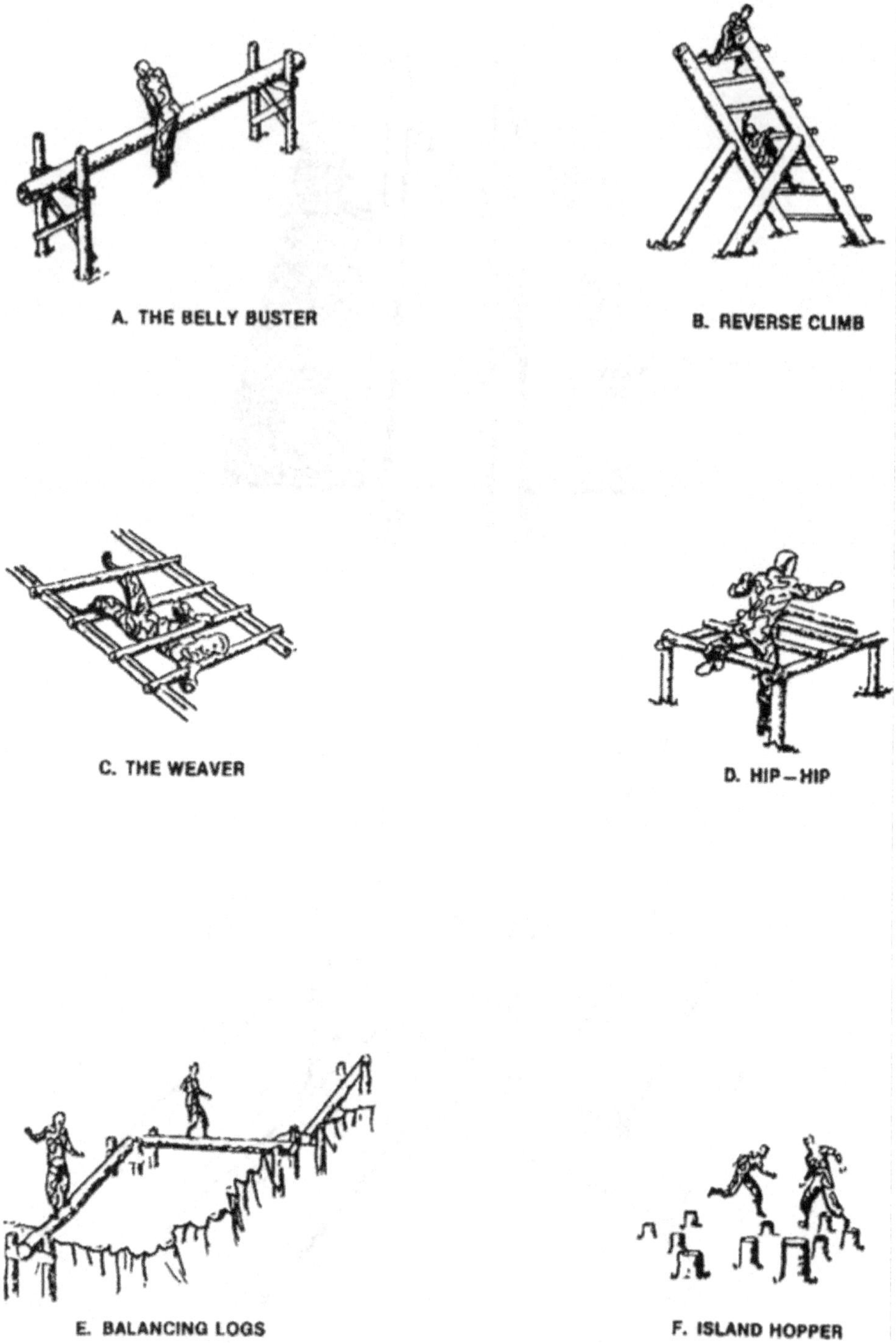

Figure 3-36.Red Group.

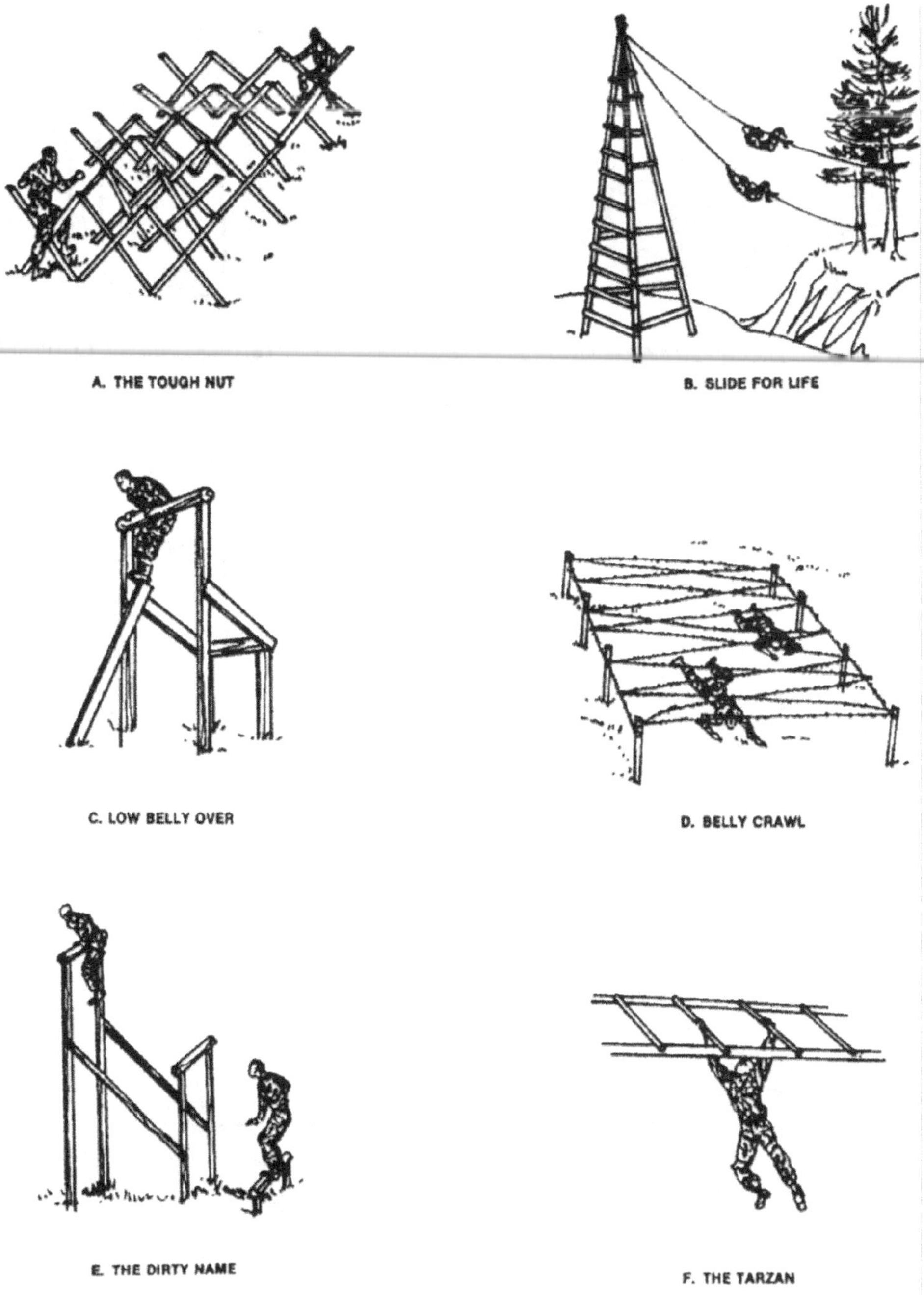

Figure 3-37.White Group.

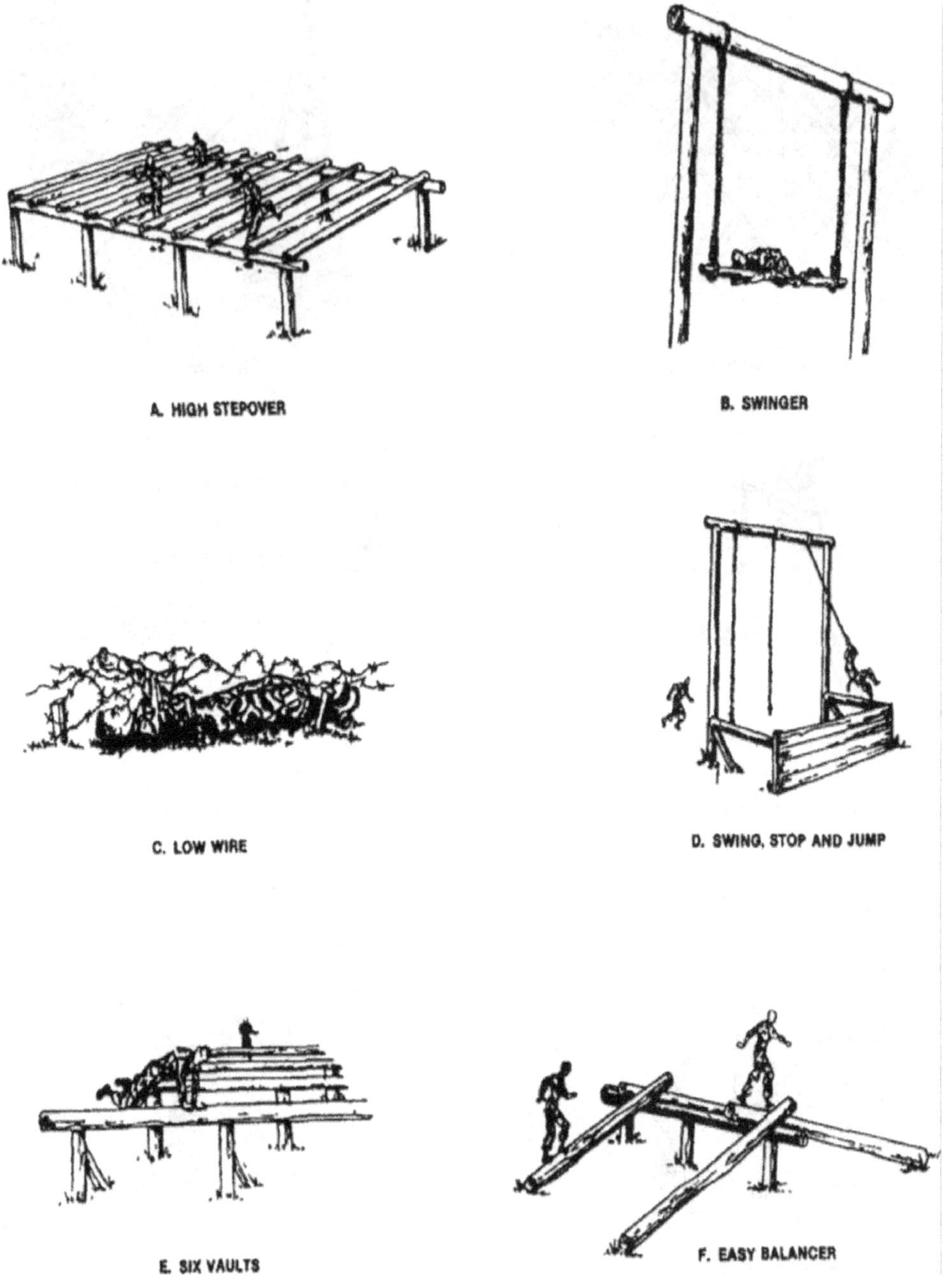

Figure 3-38.Blue Group.

A. INCLINE WALL

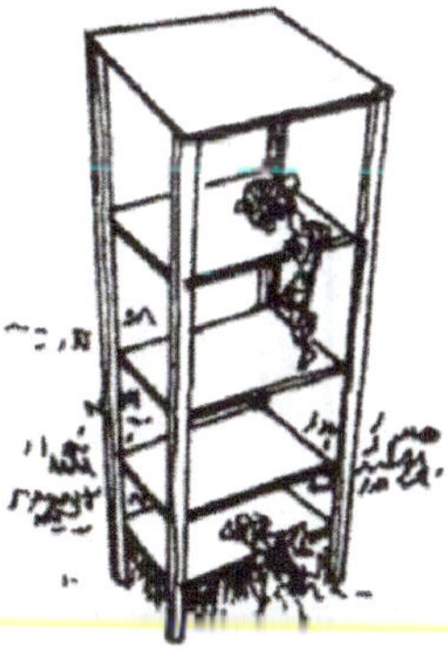

B. SKYSCRAPER

C. JUMP AND LAND

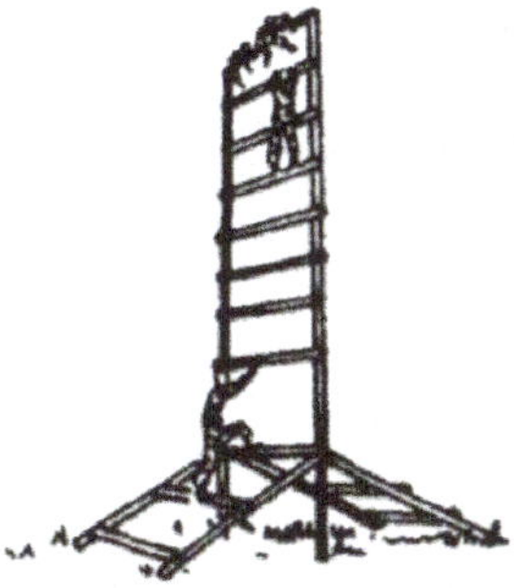

D. CONFIDENCE CLIMB

E. BELLY ROBBER

F. THE TOUGH ONE

Figure 3-39.Black Group.

Figure 3-40. Bench Conditioning, Table I.

Figure 3-41. Bench Conditioning, Table II.

PROGRESSION GUIDE

AGE GROUP	LEVEL	EXERCISES 1	2	3	4	5	6
17	A	15	18	14	15	15	250
to	B	13	16	13	13	13	235
29	C	11	14	12	11	11	[illegible]
30	A	13	14	12	13	13	200
to	B	11	13	11	11	11	185
39	C	9	12	10	9	9	165
40	A	11	11	10	11	11	150
to	B	9	10	9	9	9	135
44	C	7	9	8	7	7	120
45	A	9	8	8	9	9	100
to	B	7	7	7	7	7	90
49	C	5	6	6	5	5	80
50	A	7	6	6	7	7	75
to	B	5	5	5	5	5	70
59	C	3	4	4	3	3	60
60	A	4	5	4	4	4	50
and	B	3	4	3	3	3	40
over	C	2	3	2	2	2	30
Minutes for each exercise		2	1	1	1	2	5

Table I. Progression Guide.

Table I. Exercises.

PROGRESSION GUIDE

AGE GROUP	LEVEL	EXERCISES 1	2	3	4	5	6
17	A	17	17	17	9	19	300
to	B	15	15	15	8	17	270
29	C	13	13	13	7	15	245
30	A	15	15	15	8	17	235
to	B	13	13	13	7	15	210
39	C	11	11	11	6	13	190
40	A	13	13	13	7	15	175
to	B	11	11	11	6	13	155
44	C	9	10	9	5	11	135
45	A	11	11	11	6	13	125
to	B	9	9	9	5	11	110
49	C	7	7	7	4	9	100
50	A	9	9	9	5	11	95
to	B	7	7	7	4	9	85
59	C	5	5	5	3	7	75
60	A	6	7	7	4	9	70
and	B	5	5	5	3	7	60
over	C	4	4	4	2	5	50
Minutes for each exercise		1	1	1	1 1/2	1 1/2	6

Table II. Progression Guide.

EXERCISE 1

EXERCISE 2

EXERCISE 3

EXERCISE 4

EXERCISE 5

EXERCISE 6

Table II. Exercises.

PROGRESSION GUIDE

AGE GROUP	LEVEL	EXERCISES 1	2	3	4	5	6
17	A	10	19	19	16	10	350
to	B	9	17	17	15	9	315
29	C	8	15	15	14	8	280
30	A	9	17	17	14	9	270
to	B	8	15	15	13	8	240
39	C	7	13	13	12	7	210
40	A	8	15	15	12	8	200
to	B	7	13	13	11	7	180
44	C	6	11	11	10	6	160
45	A	7	13	13	10	7	150
to	B	6	11	11	9	6	135
49	C	5	9	9	8	5	120
50	A	6	11	11	8	6	115
to	B	5	9	9	7	5	105
59	C	4	7	7	6	4	95
60	A	5	9	9	7	5	90
and	B	4	7	7	6	4	80
over	C	3	5	5	4	3	70
Minutes for each exercise		1 1/2	1	1	1 1/2	1	6

Table III. Progression Guide.

Table III. Exercises.

PROGRESSION GUIDE

AGE GROUP	LEVEL	EXERCISES 1	2	3	4	5	6
17	A	12	9	12	24	25	400
to	B	11	8	11	22	23	380
29	C	10	7	10	21	21	360
30	A	11	8	11	23	23	305
to	B	10	7	10	21	21	290
39	C	9	6	9	20	20	275
40	A	10	7	10	20	21	225
to	B	9	6	9	18	18	215
44	C	8	5	8	16	14	205
45	A	8	6	8	16	14	175
to	B	7	5	7	14	14	165
49	C	6	4	6	12	12	155
50	A	6	5	6	13	13	135
to	B	5	4	5	11	11	130
59	C	4	3	4	10	10	120
60	A	5	4	5	10	10	100
and	B	4	3	4	9	9	95
over	C	3	2	3	8	8	90
Minutes for each exercise		1	2	1	1	1	6

Table IV. Progression Guide.

Table IV. Exercises.

PROGRESSION GUIDE

AGE GROUP	LEVEL	EXERCISES 1	2	3	4	5	6
17	A	14	13	28	14	30	450
to	B	13	12	27	13	28	430
29	C	12	11	26	12	26	410
30	A	12	12	25	12	26	350
to	B	11	11	24	11	24	330
39	C	10	10	23	10	22	310
40	A	11	11	23	11	23	250
to	B	10	10	21	10	21	240
44	C	9	9	19	9	19	230
45	A	9	9	20	9	20	200
to	B	8	8	18	8	18	190
49	C	7	7	16	7	16	180
50	A	7	7	16	7	16	170
to	B	6	6	14	6	14	155
59	C	5	5	12	5	12	140
60	A	6	6	12	6	12	115
and	B	5	5	11	5	10	110
over	C	4	4	9	4	9	105
Minutes for each exercise		2	1	1	2	1	5

Table V. Progression Guide.

Table V. Exercises.

PROGRESSION GUIDE

AGE GROUP	LEVEL	EXERCISES 1	2	3	4	5	6
17	A	17	15	32	32	35	500
to	B	16	14	30	30	33	480
29	C	15	13	28	28	31	460
30	A	15	13	30	30	31	400
to	B	14	12	28	28	29	380
39	C	13	11	26	26	27	360
40	A	13	10	27	27	27	310
to	B	12	9	25	25	25	285
44	C	11	8	23	23	23	265
45	A	11	9	23	23	23	250
to	B	10	8	21	21	21	230
49	C	9	7	19	19	19	210
50	A	9	8	19	19	19	200
to	B	8	7	17	17	17	190
59	C	7	6	15	15	15	175
60	A	8	7	15	15	17	140
and	B	7	6	13	13	15	130
over	C	5	5	10	10	12	120
Minutes for each exercise		2	1	1	1	1	6

Table VI. Progression Guide.

Table VI. Exercises.

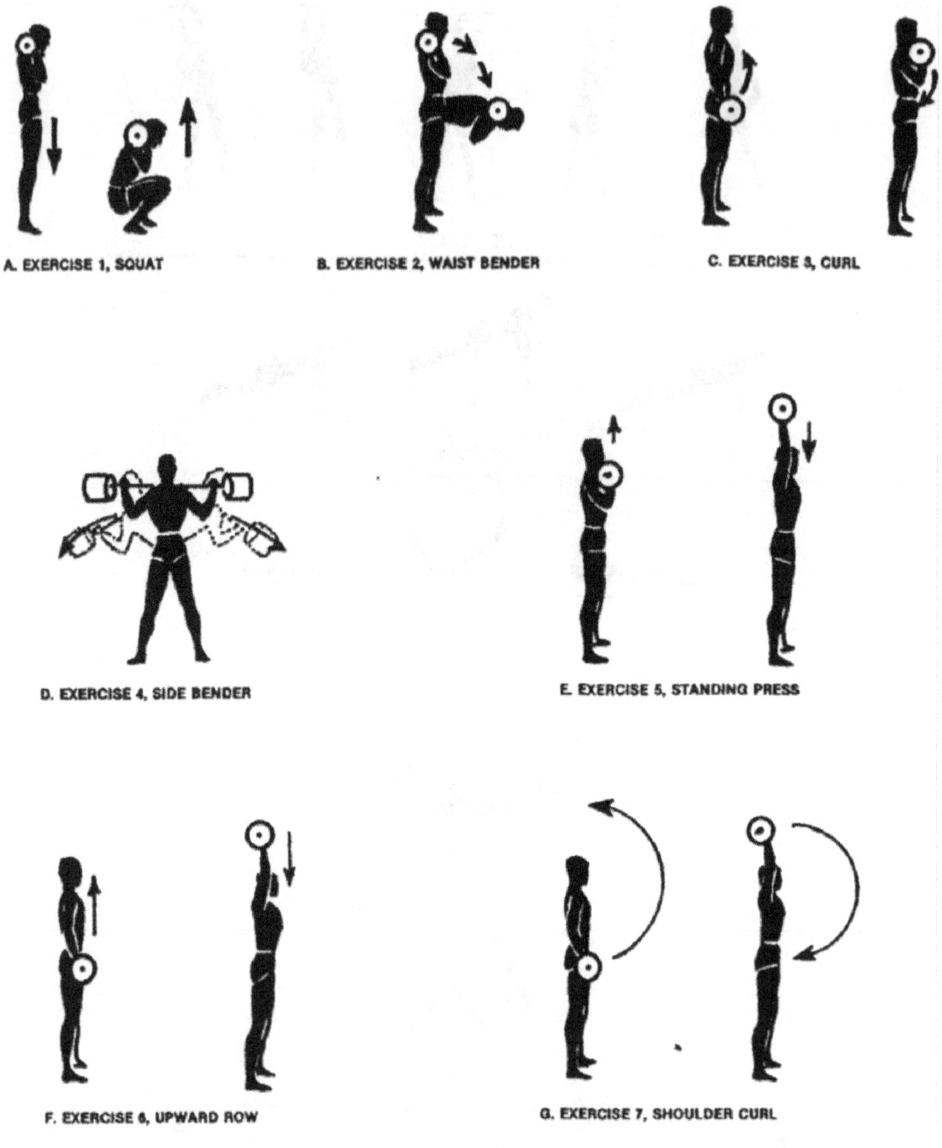

Figure 3-42. Weight Training.

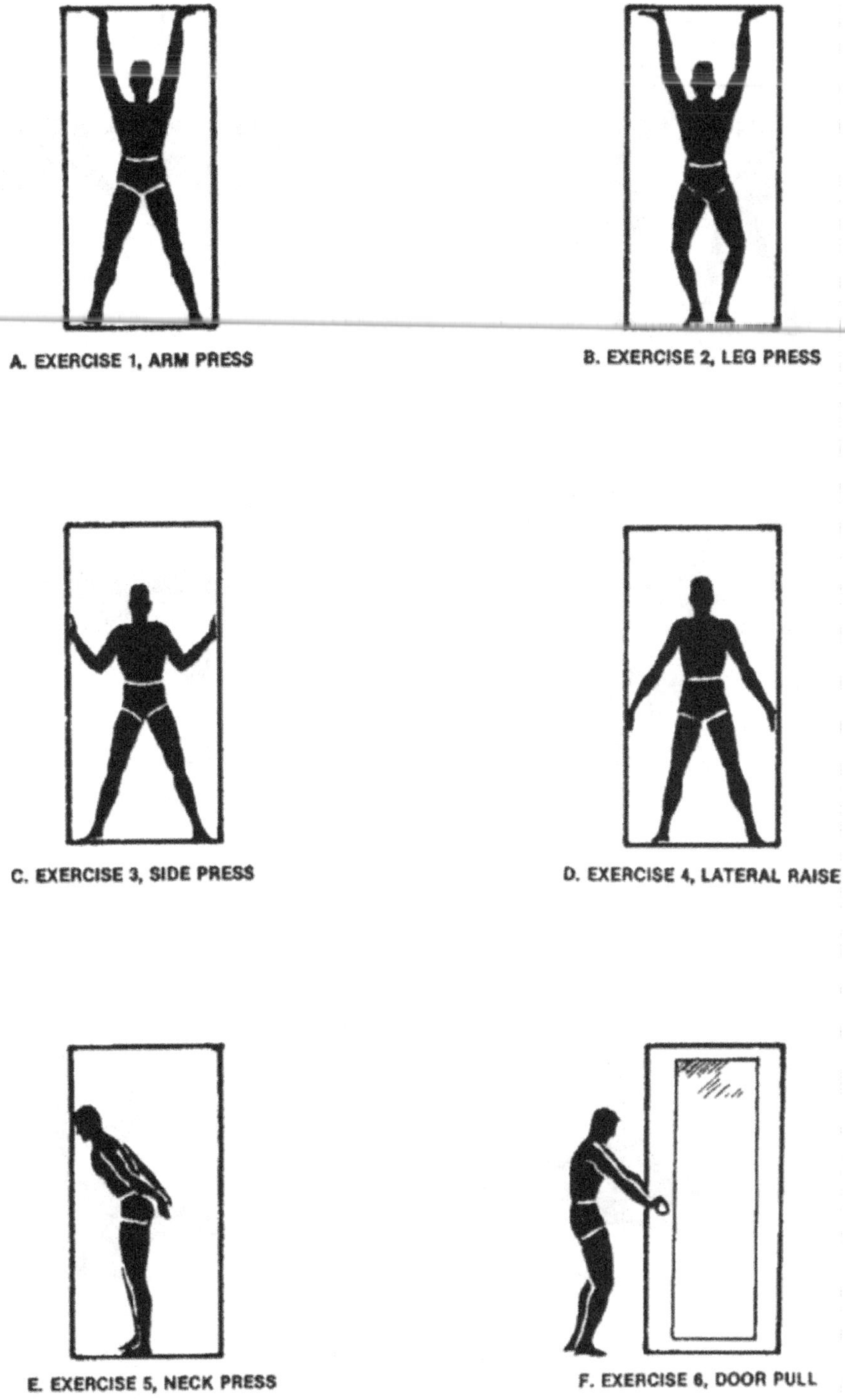

Figure 3-43. Isometric Exercises.

Figure 5-1.Combatives Table I.

Figure 5-2.Combatives Table II.

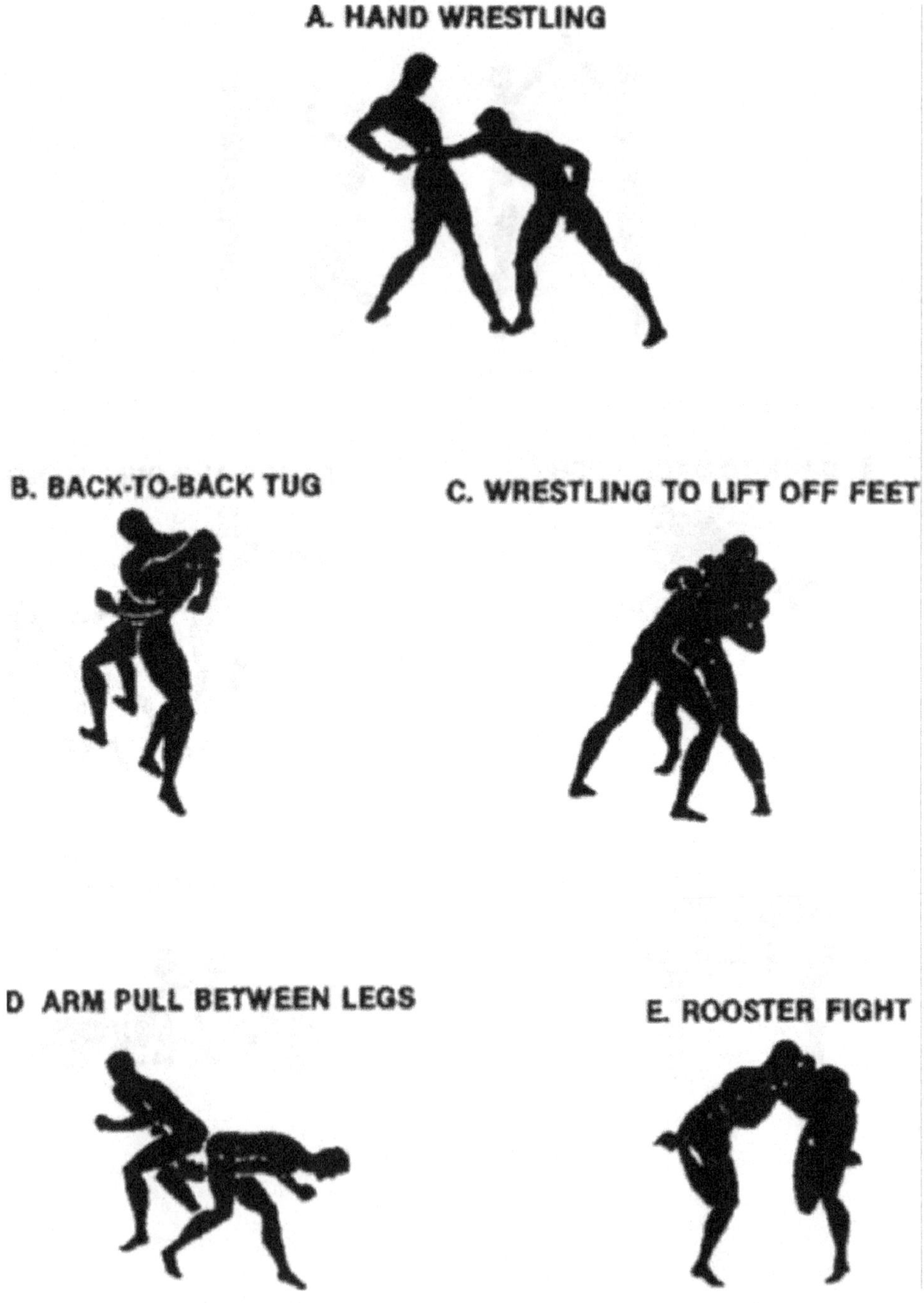

Figure. 5-3.Combatives Table III.

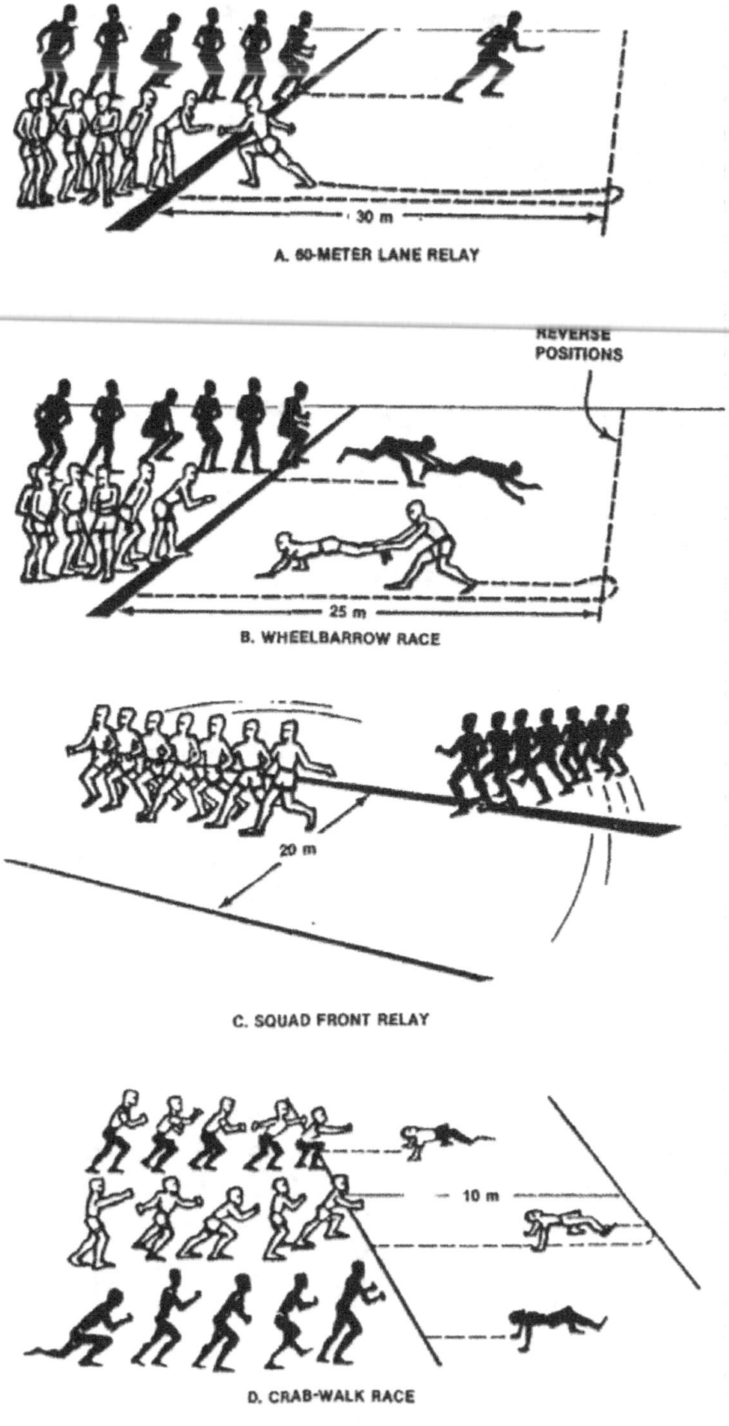

Figure 5-4.Relay Table I.

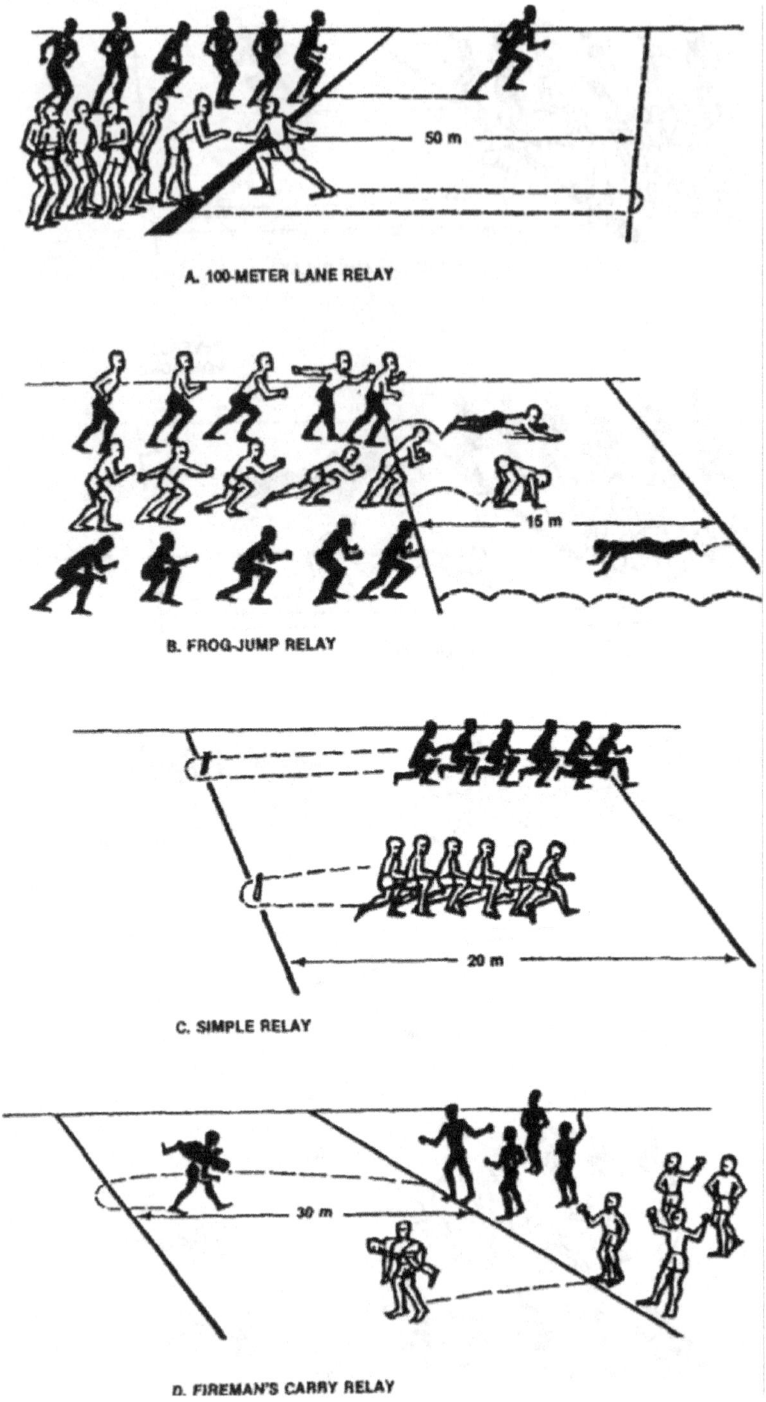

Figure 5-5.Relay Table II.

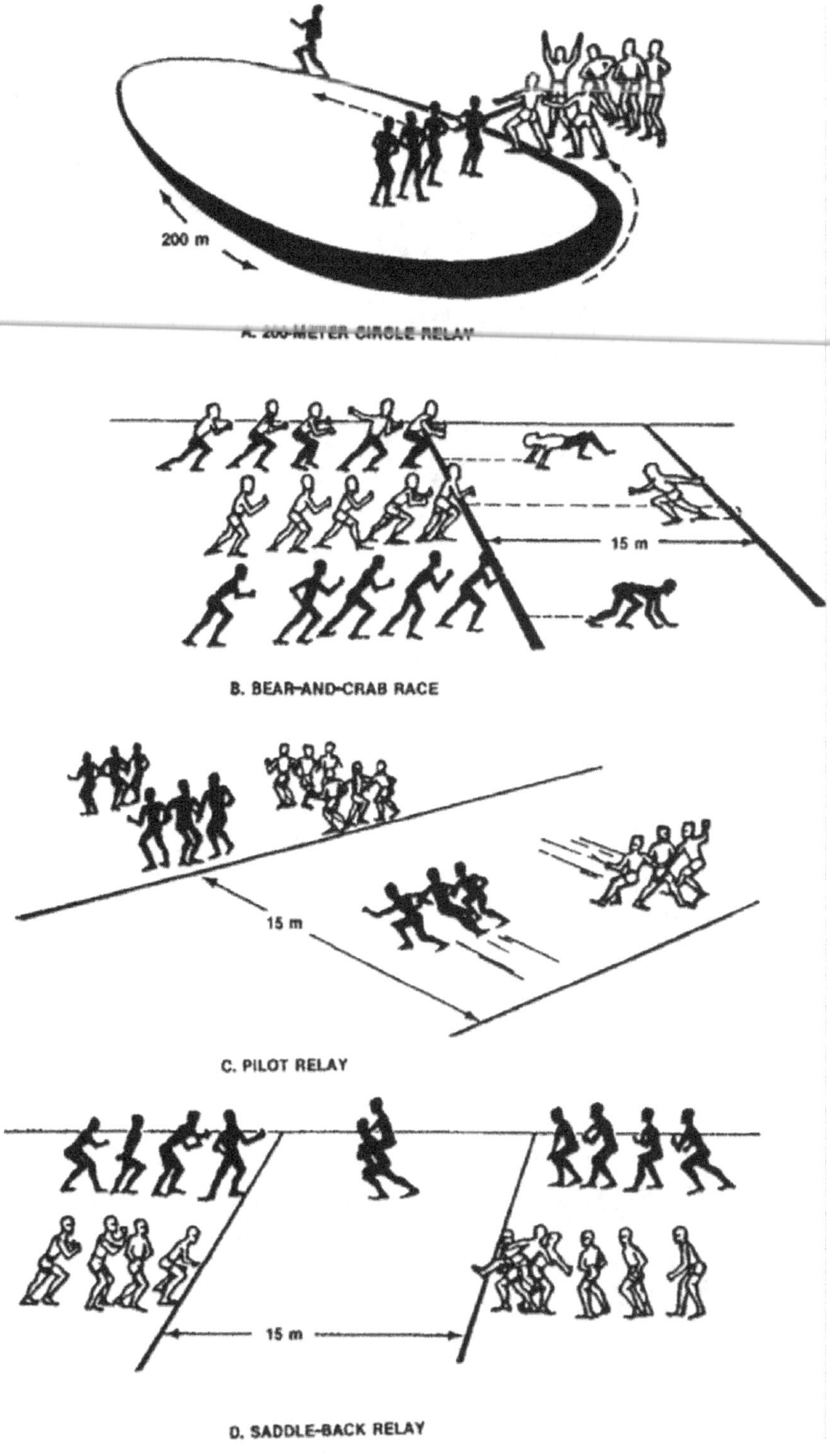

Figure 5-6.Relay Table III.

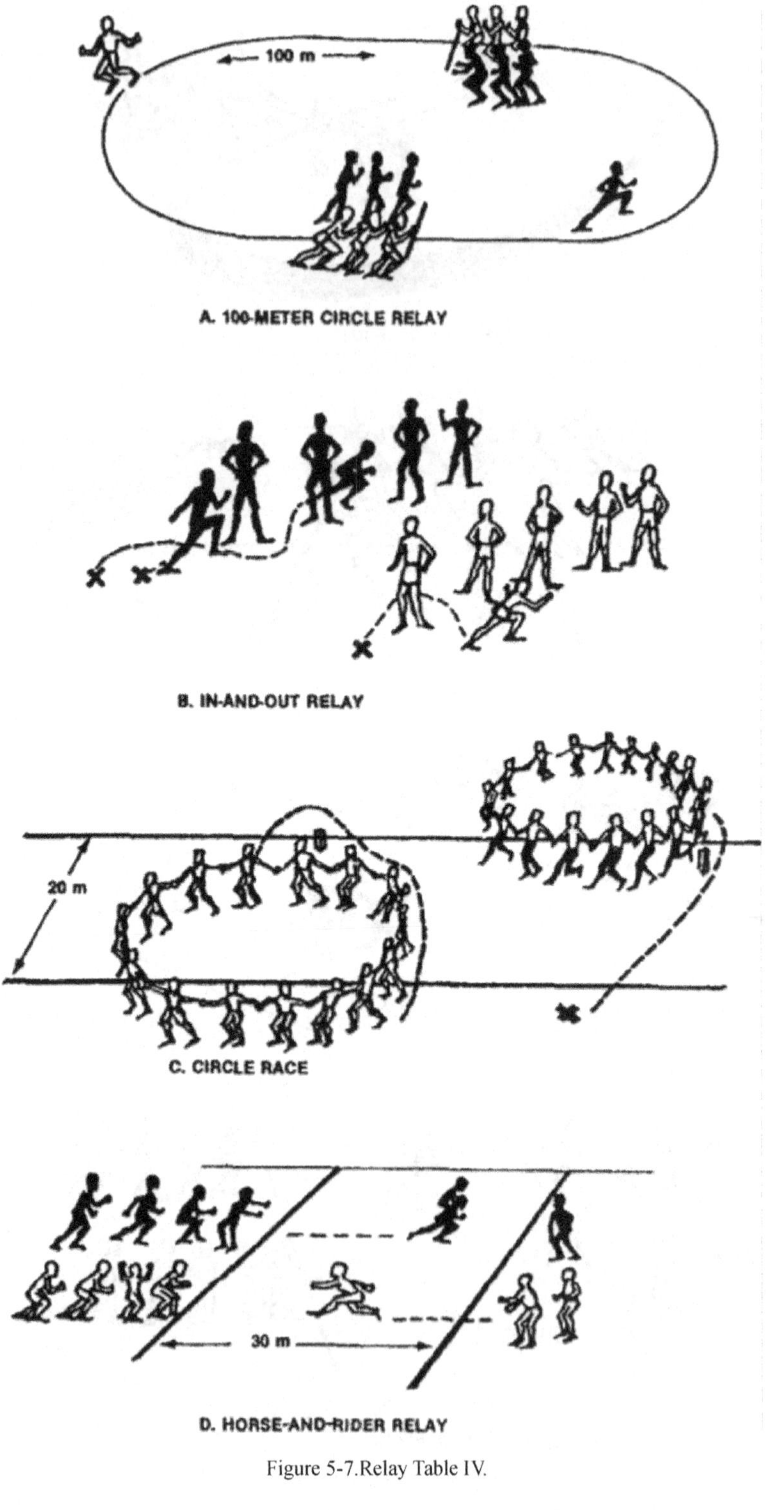

Figure 5-7.Relay Table IV.

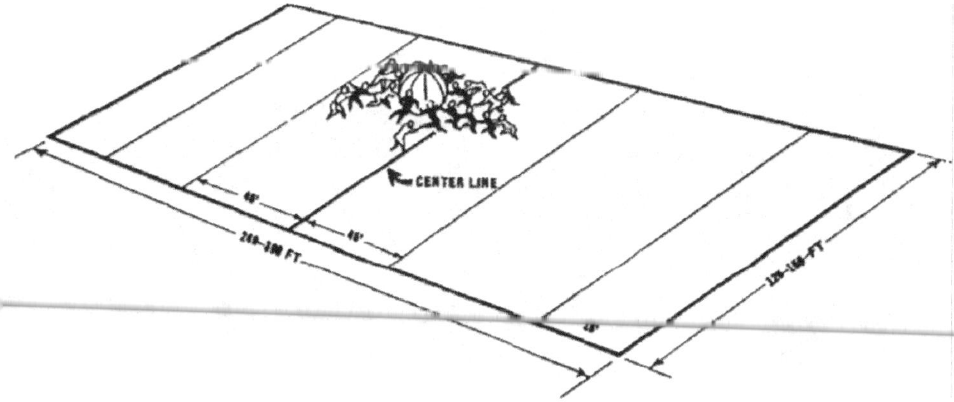

Figure 5-8.Pushball.

Figure 5-9.Line Rush.

Figure. 5-10. Human Tug-of-War.

Figure 5-11. Master of the Ring.

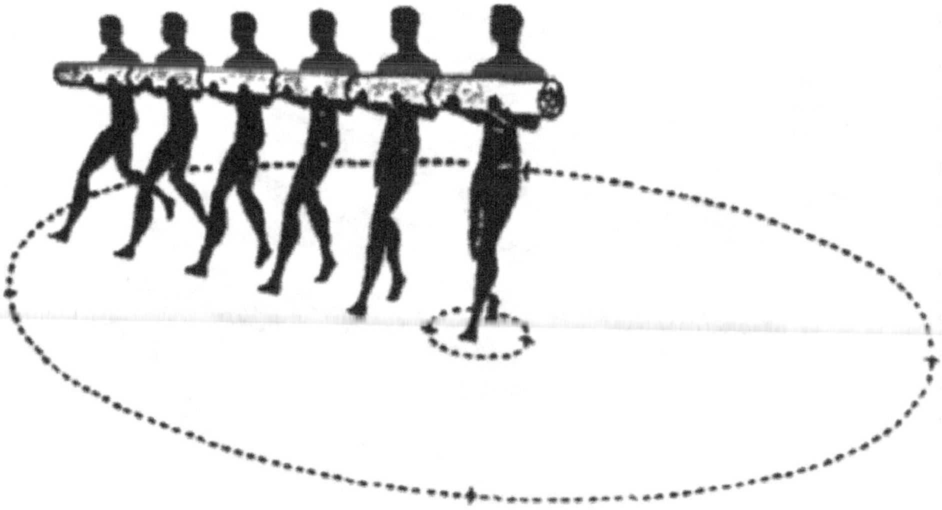

Figure 5-12.Log Pivot Circle.

Figure 5-13.Log-Rolling Race.

Figure 5-14. Prone Pushing Contest.

Figure 5-15. Shuttle Relay Race.

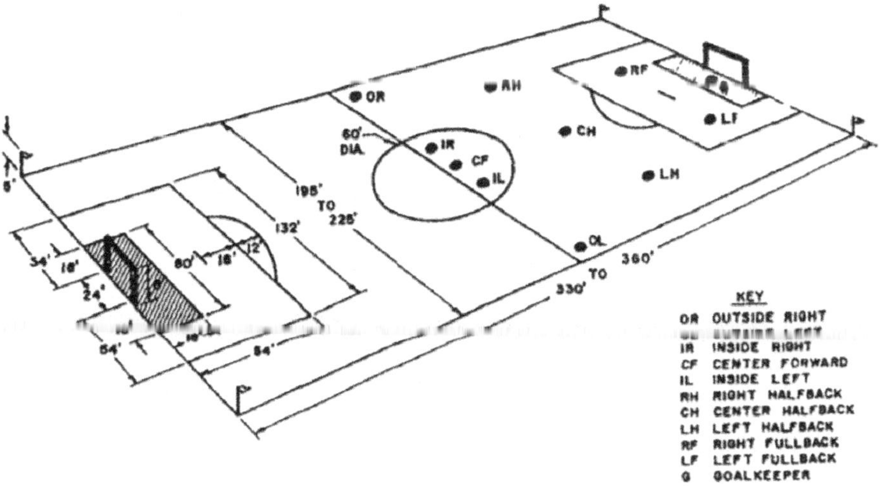

Figure 5-16. Soccer.

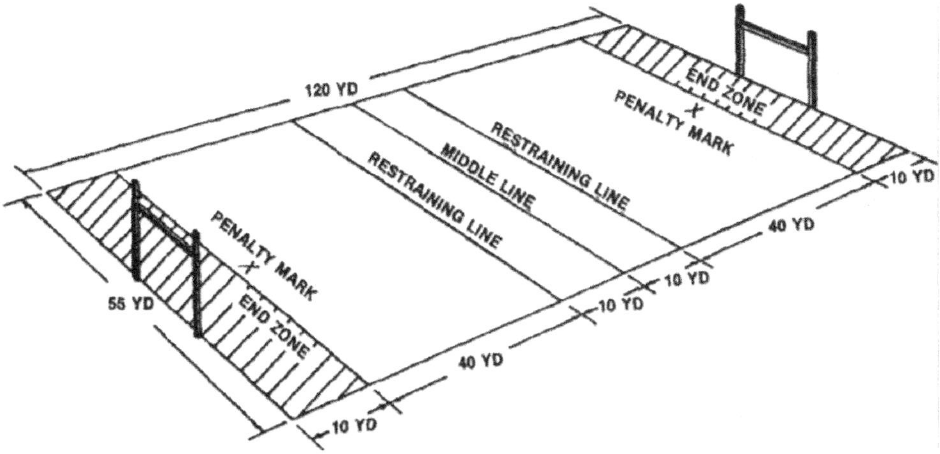

Figure 5-17.Speedball Field.

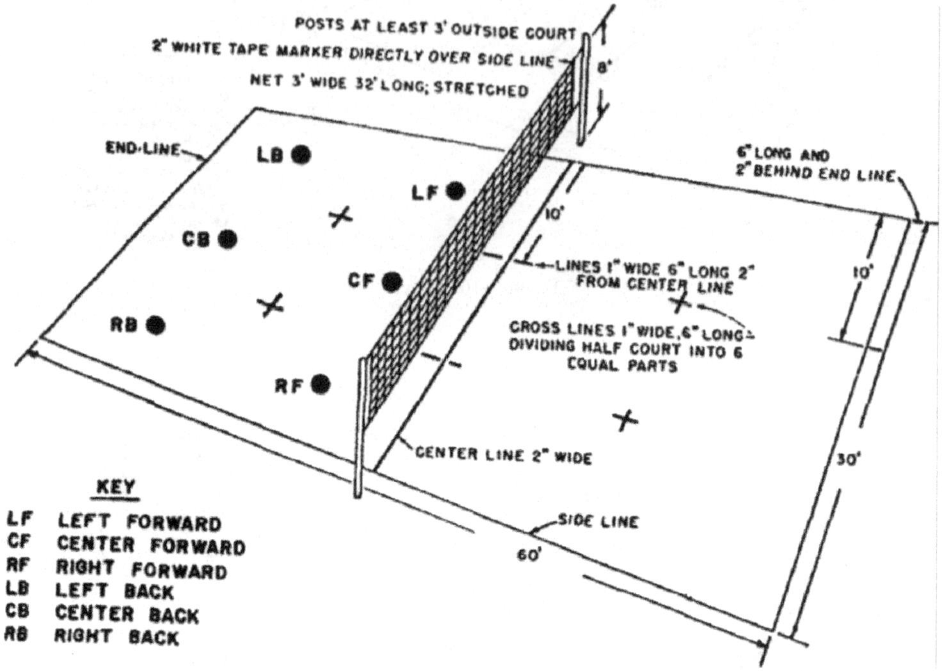

Figure 5-18. Volleyball.

PTS	UPHILL CLIMB	ROPE CLIMB	EVACUATION	FIRE AND MANEUVER	FORCED MARCH
100	1:20-BELOW	0:10-BELOW	0:25-BELOW	1:20-BELOW	26:00-BELOW
99	1:21-1:22	0:11		1 21-1:22	26:01-26:12
98	1:23-1:24	0:12	0:26	1:23-1:24	26:13-26:24
97	1:25-1:26	0:13		1:25-1:26	26:25-26:36
96	1:27-1:28	0:14	0:27	1:27-1:28	26:37-26:48
95	1:29-1:30	0:15		1:29-1:30	26:49-27:00
94	1:31-1:32	0:16	0:28	1:31-1:32	27:01-27:12
93	1:33-1:34	0:17		1 33-1.34	27:13-27:24
92	1:35-1:36	0:18	0:29	1:35-1.36	27:25-27:36
91	1:37-1:38	0:19		1:37-1:38	27:37-27:48
90	1:39-1:40	0:20	0:30	1:39-1:40	27:49-28:00
89	1:41-1:42	0:21	0:31	1:41-1:42	28:01-28:12
83	1:43-1:44	0:22	0:32	1:43-1:44	28:13-28:24
87	1:45-1:46	0:23	0:33	1:45-1:46	28:25-28:36
86	1:47-1:48	0:24	0:34	1.47-1:48	28:37-28:48
85	1:49-1:50	0:25	0:35	1.49-1:50	28:49-29:00
84	1:51-1:52	0:26	0:36	1:51-1:52	29:01-29:12
83	1:53-1:54	0:27	0.37	1:53-1:54	29:13-29:24
82	1:55-1:56	0:28	0:38	1:55-1:56	29:25-29:36
81	1:57-1:58	0:29	0:39	1:58-1:58	29:37-29:48
80	1:59-2.00	0:30	0:40	1:59-2:00	29:49-30:00
79	2.01-2:02	0:31	0:41	2:01-2:02	30:01-30:12
78	2:03-2:04	0:32	0:42	2:03-2.04	30:13-30:24
77	2:05-2:06	0.33	0:43	2:05-2:06	30:25-30:36
76	2:07-2:08	0.34	0.44	2:07-2:08	30:37-30:48
75	2:09-2:10	0:35	0:45	2:09-2:10	30:49-31:00
74	2:11-2:12	0:36	0:46	2:11-2 12	31.01-31:12
73	2:13-2:14	0:37	0.47	2.13-2:14	31:13-31:24
72	2:15-2:16	0:38	0:48	2.15-2:16	31:25-31:36
71	2:17-2:18	0:39	0:49	2:17-2:18	31:37-31:48
70	2:19-2:20	0:40	0:50	2:19-2.20	31:49-32.00
69	2.21-2:22	0:41	0:51	2:21-2:22	32:01-32:12
68	2:23-2:24	0:42	0:52	2:23-2:24	32:13-32:24
67	2:25-2.26	0:43	0:53	2:25-2:26	32:25-32:36
66	2:27-2:28	0:44	0:54	2.27-2:28	32.37-32:48
65	2:29-2:30	0.45	0.55	2.20-2 30	32.49-33:00

Figure 6-1.Physical Readiness Test Scoring Table for Male.

	EXAMPLE 1	EXAMPLE 2	EXAMPLE 3
Event #1	100 pts	100 pts	99 pts
Event #2	100 pts	99 pts	86 pts
Event #3	85 pts	75 pts (F)	80 pts
Event #4	90 pts	79 pts (F)	100 pts
Event #5	79 pts	85 pts	90 pts
Pts	454 pts	438 pts	455 pts
Final Score	79.0% (F)	69.0% (F)	91.0%
(F) = Failure			

Figure 6-2.Scoring Examples.

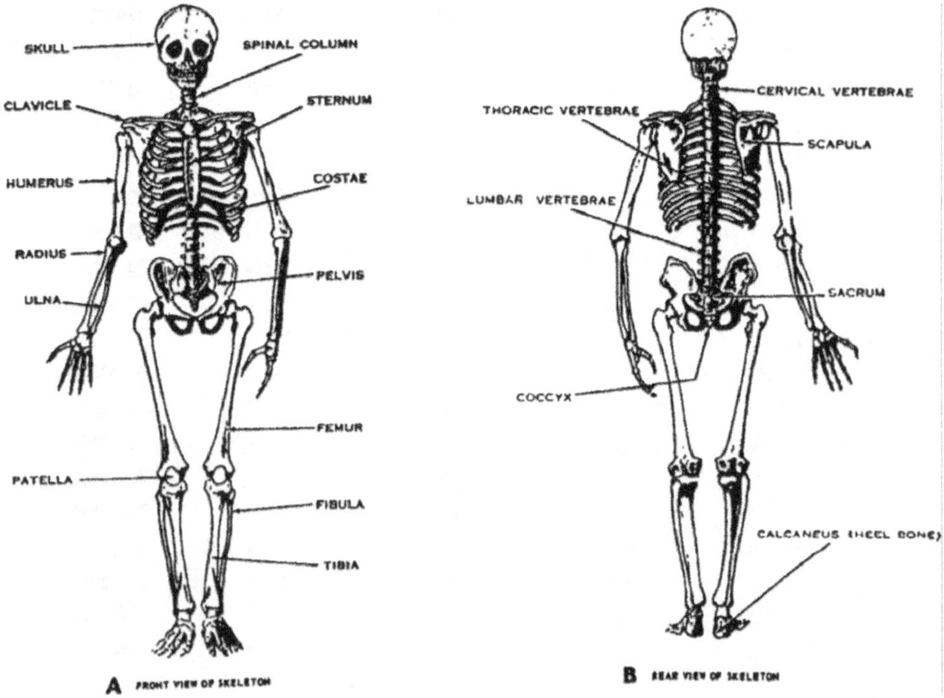

Figure 7-1.Front and Rear Views of Skeleton.

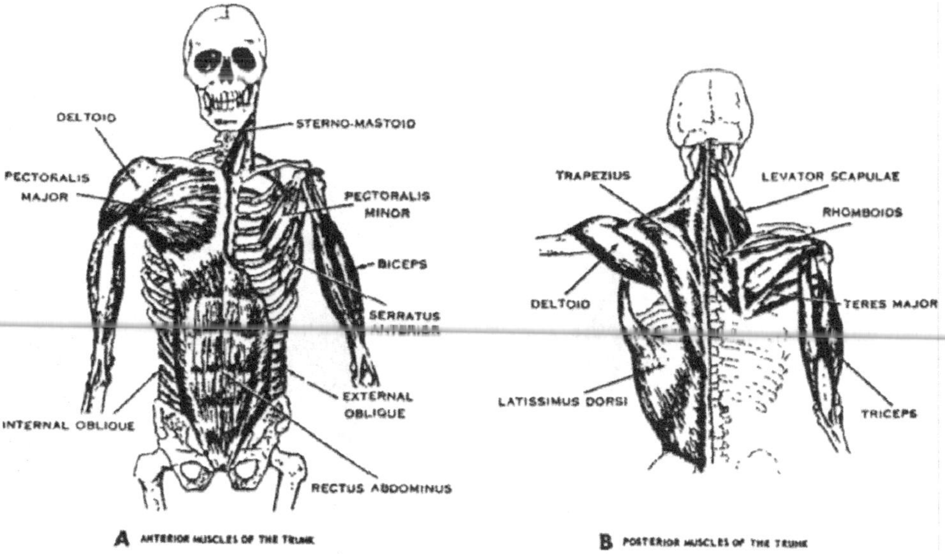

Figure 7-2.Muscles of the Trunk.

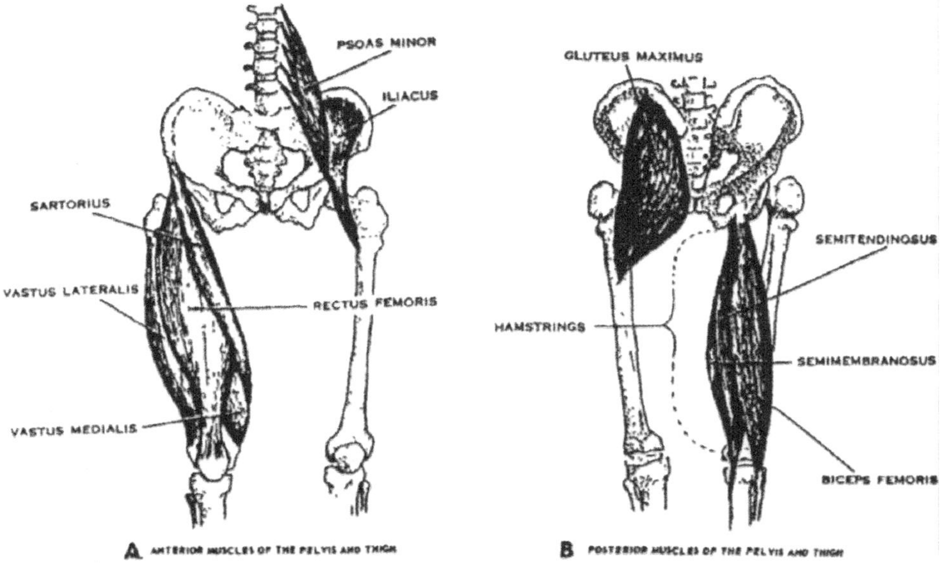

Figure 7-3.Anterior and Posterior Muscles of the Pelvisand Thigh.

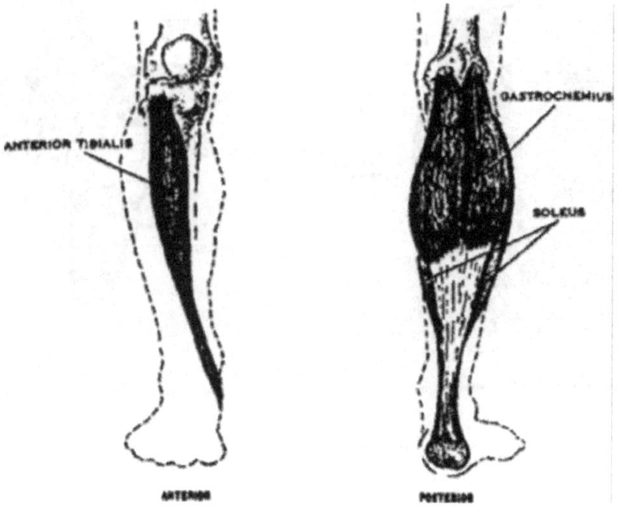

Figure 7-4. Muscles of Lower Leg.

Figure 7-5.Circulation of Blood from Heart to the Body.

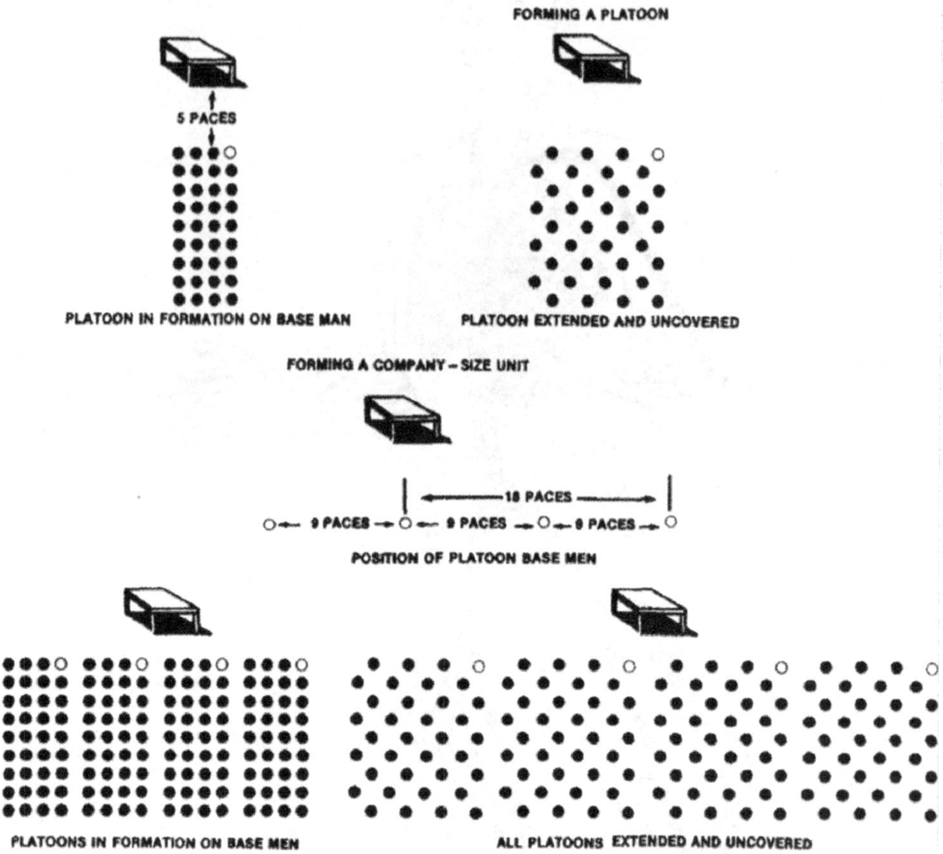

Figure A-1. Forming the Extended Rectangular Formation.

Figure A-2.Circle Formation.

Table of Contents

Chapter 1. The M9 Service Pistol

Chapter 2. Weapons Handling

Chapter 3. Fundamentals of Pistol Marksmanship

Chapter 4. Pistol Firing Positions and Grip

Chapter 5. Use of Cover and Concealment

Chapter 6. Presentation of the M9 Service Pistol

Chapter 7. Pistol Engagement Techniques

CHAPTER 1
THE M9 SERVICE PISTOL

The Marine Corps' uses the M9 service pistol as a defensive weapon. It is a semiautomatic, magazine-fed, recoil-operated, double-action pistol. The M9 service pistol's magazine holds 15 rounds and is chambered with a 9-mm ball (NATO M882 round). The M9 service pistol has a maximum effective range of 50 meters (54.7 yards).

Note: The procedures in this chapter are written for right-handed Marines; left-handed Marines reverse instructions as necessary.

Functional Capabilities

When the M9 service pistol's safety lever is moved to the firing position, it can be fired in either a single-action or double-action mode. It is designed to fire one round each time the trigger is pulled. When the last round is fired, the slide automatically locks to the rear.

Single-Action Mode

The single-action mode allows the pistol to be fired when the hammer is cocked; single action requires the hammer to be cocked to the rear before the trigger is pulled. The hammer can be manually cocked or mechanically cocked. The hammer is mechanically cocked after the first shot is fired. See figure 1-1.

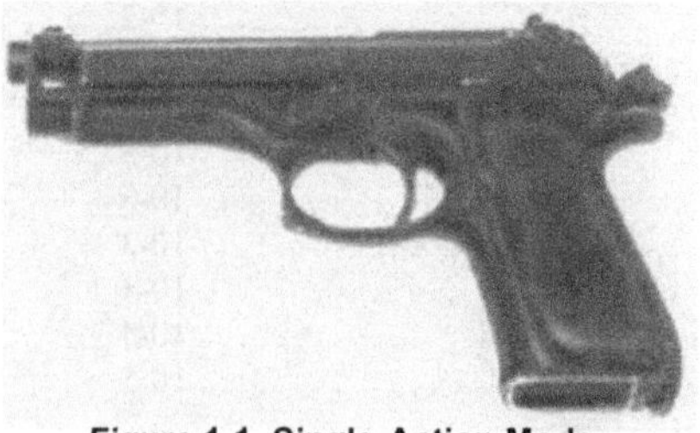

Figure 1-1. Single-Action Mode.

Double-Action Mode

The double-action mode causes the hammer to move to the rear as the trigger is being pulled. See figure 1-2.

Figure 1-2. Double-Action Mode.

Nomenclature

See figure 1-3 on page 1-2.

Major Components

The M9's major components consist of the slide assembly, barrel assembly, and receiver. See figure 1-4 on page 1-3.

Slide Assembly

The slide assembly houses the firing pin, firing pin block, striker, extractor, and loaded chamber indicator, and it cocks the hammer during recoil.

Extractor

The extractor pulls the brass from the chamber after the round is fired.

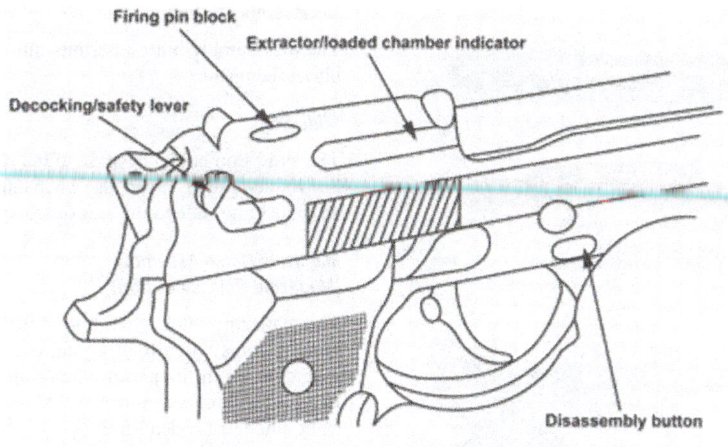

Right Side View

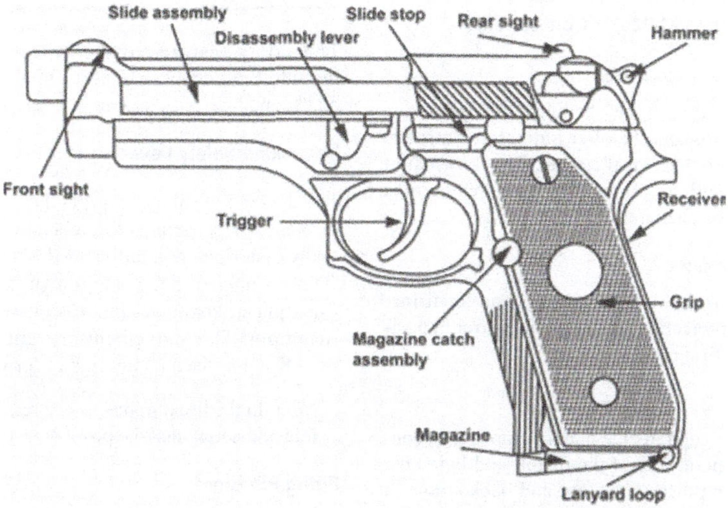

Left Side View

Figure 1-3. M9 Service Pistol.

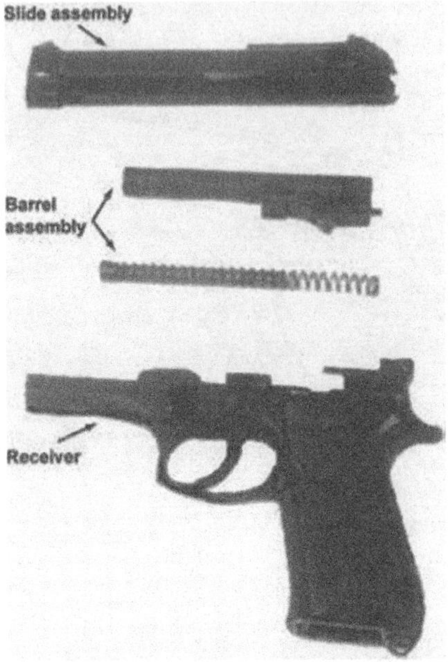

Figure 1-4. Major Components.

Loaded Chamber Indicator

When a round is in the chamber, the upper surface of the loaded chamber indicator protrudes from the right side of the slide. This protrusion can be felt with the finger, verifying that there is a round in the chamber.

Barrel Assembly

The barrel assembly houses the round for firing, directs the projectile, and locks the barrel in position during firing.

Receiver

The receiver supports the major components, controls the functioning of the pistol, and holds the magazine in place. The front and back straps of the grip are grooved vertically to ensure that the hand does not slip while firing. The receiver consists of the disassembly button, slide stop, and magazine catch assembly.

Disassembly Button

The disassembly button permits quick disassembly of the pistol.

Slide Stop

The slide stop holds the slide to the rear after the last round is fired. It can also be manually operated to lock the slide to the rear or release the slide.

Magazine Catch Assembly (Magazine Release Button)

The magazine catch assembly secures the magazine in place when loading, and it releases the magazine from the pistol when unloading. The magazine catch assembly is designed for both right- and left-handed Marines.

> *Note:* Reversal of the magazine catch assembly for left-handed Marines can be performed by a qualified armorer.

Safety Features

The safety features of the M9 service pistol include the decocking/safety lever, firing pin block, and half-cock notch.

Decocking/Safety Lever

The decocking/safety lever, commonly referred to as the safety, permits safe operation of the pistol by both right- and left-handed Marines. As the safety is moved to the safe (down) position, the firing pin striker moves out of alignment with the firing pin. This movement prevents the pistol from firing as the hammer moves forward.

> *Note:* In the fire (up) position, a red dot is visible, indicating that the pistol is ready to fire.

Firing Pin Block

The firing pin block rests in the firing pin notch and prevents movement of the firing pin until the trigger is pulled. As the trigger is pulled, the firing pin block moves up and out of the firing pin notch. This movement allows a round to be fired when the hammer strikes the firing pin.

Half-Cock Notch

The half-cock notch stops the forward movement of the hammer during a mechanical failure.

Cycle of Operation

There are eight steps in the cycle of operation for the M9 service pistol.

Firing

Once the safety is off and the trigger is pulled to the rear, the hammer falls on the firing pin, which strikes the primer and ignites the round. See figure 1-5.

Figure 1-5. Firing.

Unlocking

As the slide assembly moves to the rear, the locking block rotates out of the notches in the slide. See figure 1-6.

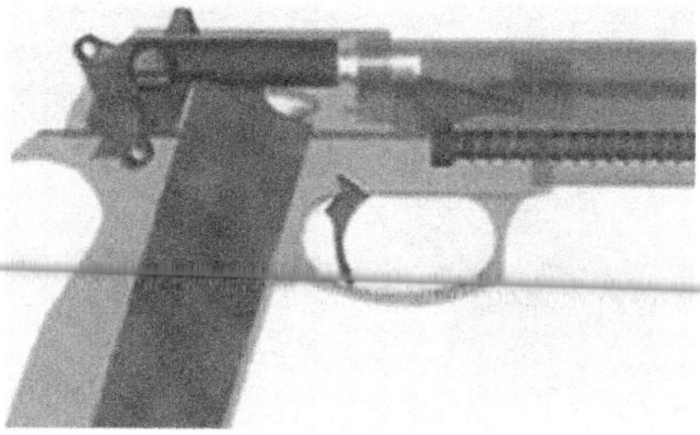

Figure 1-6. Unlocking.

Extracting

As the slide moves rearward, the extractor withdraws the cartridge case out of the chamber. See figure 1-7.

Figure 1-7. Extracting.

Ejecting

As the face of the slide passes over the ejector, the case strikes the ejector and it is knocked upward and outward through the ejection port. See figure 1-8.

Figure 1-8. Ejecting.

Cocking

As the slide moves rearward, the hammer is pushed back, allowing the sear to engage the hammer hooks, cock the hammer to the rear, and place the pistol in the single-action mode. See figure 1-9.

Figure 1-9. Cocking.

Feeding

The slide starts forward, pushed by the recoil spring. The face of the slide makes contact with the cartridge at the top of the magazine, stripping it from the magazine and pushing it toward the chamber. See figure 1-10.

Figure 1-10. Feeding.

Chambering

As the slide continues forward, it pushes the cartridge into the chamber. See figure 1-11.

Figure 1-11. Chambering.

Locking

As the slide assembly continues to move forward, the locking block lugs move into the locking block recesses on the right and left sides of the slide. See figure 1-12 on page 1-6.

Figure 1-12. Locking.

Ammunition

The only ammunition authorized for the M9 service pistol is the NATO M882 9-mm ball. Dummy ammunition can be used during training (a dummy round has a hole drilled in its side and contains no primer). See figure 1-13.

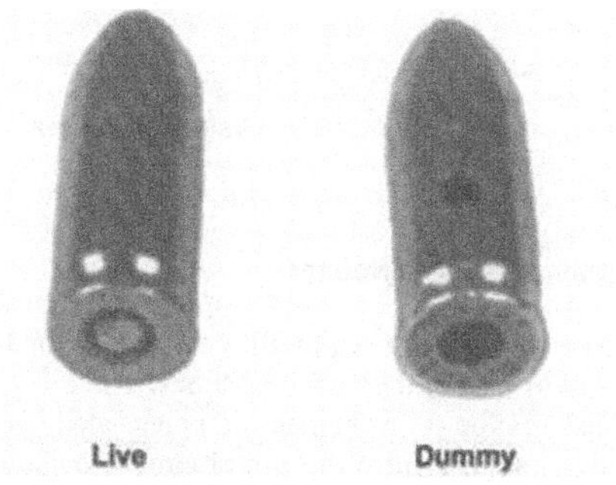

Figure 1-13. M9 Service Pistol Ammunition.

Do not open ammunition containers until the ammunition is to be used. Ammunition must be maintained in a high state of readiness. To care for ammunition—

- Keep ammunition dry and clean. If ammunition gets wet or dirty, wipe it off with a clean dry cloth.
- Wipe off light corrosion as soon as it is discovered. Never use ammunition that is heavily corroded, dented, or has the projectile pushed in.
- Do not expose ammunition to direct sunlight for long periods of time.
- Do not oil or grease ammunition. Dust or other abrasives can collect on greasy ammunition and may cause damage to the operating parts of the pistol. Oiled cartridges also produce excessive chamber pressure.

Wearing of the M9 Service Pistol's Gear

The proper placement of pistol gear helps ensure safety and aids the Marine in effectively handling and employing the pistol.

M12 Holster

The M12 holster consists of the holster, removable holster flap, and metal retaining clip. To check for proper placement of the holster, allow the right arm to hang freely. The holster should be slightly in front of the arm to permit easy access to the pistol upon presentation from the holster. See figure 1-14.

Note: In most cases, the holster is issued with the holster flap installed for a right-handed Marine. To convert the holster for a left-handed Marine, remove the metal retaining clip and install the clip on the opposite side of the holster.

Figure 1-14. The M12 Holster and Ammunition Pocket.

M1 Ammunition Pocket

The ammunition pocket attaches to the cartridge belt on the side opposite the holster in a position that best permits ready access for reloads. The magazine is stored in the ammunition pocket with the rounds down and pointed inboard.

M7 Shoulder Holster

The M7 shoulder holster consists of a holster with a thumb snap closure, shoulder strap, chest strap, and a belt retaining loop. The holster is positioned on the left side of the chest to provide easy and quick access with the right hand. See figure 1-15. The M7 holster comes fully assembled and has adjustable straps to accommodate each Marine's body size. The holster is available for right-handed Marines only, therefore, a left-handed Marine has to withdraw the pistol from the holster with the right hand and then transfer the pistol to the left hand before firing. (See chap. 8 for transferring the pistol from one hand to the other.) When the shoulder holster is worn properly—

Figure 1-15. M7 Shoulder Holster.

- The shoulder strap lays flat across the left shoulder with the shoulder pad directly on top of the shoulder.
- The chest strap attaches to the D-ring at the top of the holster and runs diagonally across the chest, underneath the right arm, and around the back where it attaches to the end of the shoulder strap.
- The belt retaining loop is at the bottom of the holster and attaches to the belt to stabilize the holster's position.
- The ammunition pocket attaches to the chest strap directly underneath the right arm.

Assault Holster

Some Marines (i.e., Marine security force, direct action platoon, and military police) are required to carry the assault holster. This holster has a retention strap that fastens over the top of the holster to retain the pistol. This type of holster generally has a thumb break on the retention strap that is disengaged to access the pistol. See figure 1-16.

Figure 1-16. Assault Holster.

Concealed Pistol Holster

Some Marines are required to carry a concealed pistol as part of their official duties. The primary consideration for placement of a concealed pistol holster is to ensure the pistol cannot be seen; therefore, the Marine must consider the type of clothing to be worn. A secondary consideration is to place the holster so the pistol can be presented easily. Typically, the best position for a concealed holster is just behind the strong side hip. This position best conceals the pistol while allowing it to be presented quickly. Another placement choice is in a shoulder holster, placing the pistol just under the weak side arm. See figures 1-17 through 1-20 on pages 1-8 and 1-9.

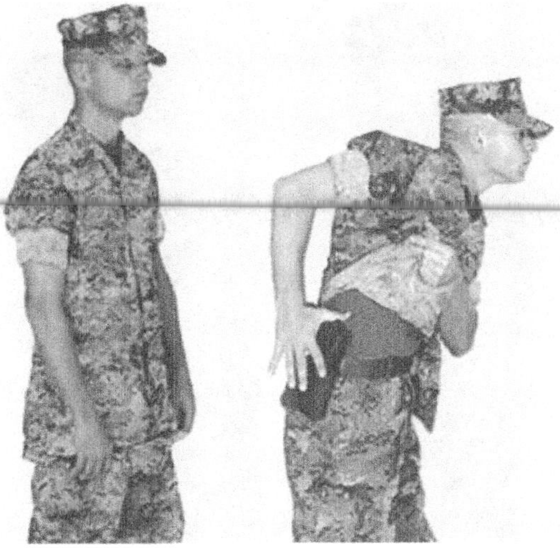

Figure 1-17. Wearing of the Concealed Pistol Holster—Utilities.

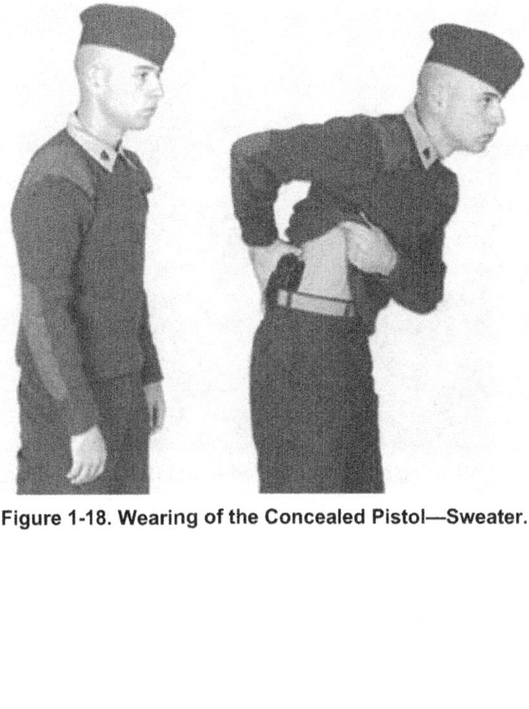

Figure 1-18. Wearing of the Concealed Pistol—Sweater.

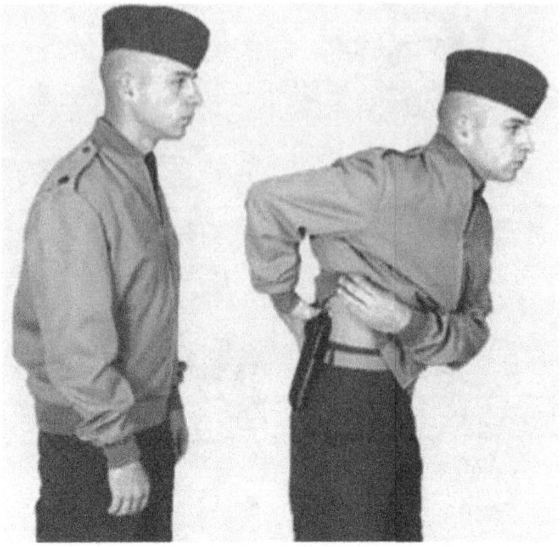

Figure 1-19. Wearing of the Concealed Pistol Holster—Jacket.

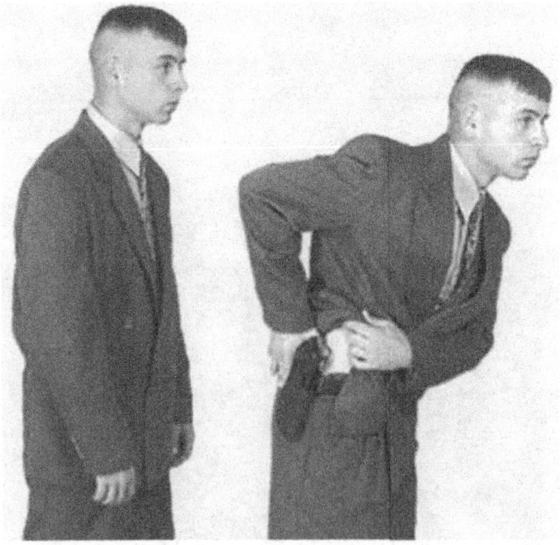

Figure 1-20. Wearing of the Concealed Pistol Holster—Civilian Attire.

Lanyard

The lanyard aids in pistol retention. It is issued in three sizes and is adjustable. The lanyard consists of a fabric cord, two cylindrical slip rings, and a metal clip that attaches the lanyard to the pistol's lanyard loop. See figure 1-21.

Figure 1-21. M9 Service Pistol (With Lanyard).

To don the lanyard—

- Adjust the slip rings so they are positioned flush with the lanyard's base (clip end).
- Place the right arm through the loop and place the loop over the head, resting on the left shoulder.
- Attach metal clip to the pistol's lanyard loop.
- Place the pistol in the holster.
- Use the left hand to hold the bottom slip ring against the base of the lanyard. Use the right hand to slide the top slip ring upward to position the loop of the lanyard under the arm. The lanyard should fit snugly against the body, but not restrict the Marine's movements.
- Tuck any excess cord behind the holster.
- Ensure that the lanyard is adjusted properly by removing the pistol from the holster and fully extending the right arm. The lanyard should be taut. Adjust as necessary.

  Note: Re-adjust the lanyard if any equipment changes are made (e.g., flak jacket).

Firing the M9 Service Pistol While Wearing Gloves

Not all combat engagements occur during ideal weather conditions. During cold weather, the Marine may find it necessary to wear gloves to protect the fingers from frostbite and help prevent stiffening of the hands. Gloves may also be worn in mission-oriented protective posture conditions. Gloves provide protection to the hands, however, they also may interfere with the Marine's ability to engage targets effectively. The added bulk of the gloves may affect the Marine's ability to manipulate the safety, magazine release button, magazine, hammer, and slide stop/release. For example, the Marine may need to exert more pressure to engage the magazine release or slide release buttons to compensate for the thickness of the gloves.

The principles of target engagement (see chap. 7) do not change while wearing gloves, however, the specific ability to manipulate and control the trigger is greatly affected by the thickness of the gloves around the fingers. Wearing gloves reduce the Marine's ability to feel, which makes it difficult to apply trigger control when firing. The Marine may find that more pressure than normal must be applied with the trigger finger just to establish initial contact with the trigger. Once the Marine can "feel" the trigger through the gloves, then the pressure required to fire a shot can be applied. This action may increase the chances of firing the pistol prematurely due to excessive pressure on the trigger. Dry firing while wearing gloves allows the Marine to learn how to apply trigger control consistently and determines how much pressure is needed to effectively fire a shot.

While wearing gloves, the Marine may find it difficult to fire the pistol in the double-action mode due to the position of the trigger and the limited amount of space between the trigger and the trigger guard. Therefore, if the situation permits, the Marine may wish to thumbcock the pistol to fire in single-action mode. In single-action mode, there is more space between the trigger guard and the trigger, making it easier to position the finger on the trigger. However, the Marine's ability to thumbcock the pistol may also be hindered by the gloves' bulk. Therefore, to thumbcock the pistol while wearing gloves, the Marine may perform one of the following methods:

- For method one, loop a section of 550 cord (approximately 2 inches) through the loop located on the top rear portion of the hammer. The length of the cord should not interfere with the pistol's cycle of operation or with the Marine's ability to establish sight alignment. Once the cord is attached to the hammer, take the pistol off safe and pull downward on the cord to cock the hammer. See figure 1-22.

Figure 1-22. Cocking the Pistol with 550 Cord.

- For method two, take the pistol off safe, rotate the pistol inboard, and place the top of the hammer against a secure surface (e.g., cartridge belt, table top, heel of boot). Apply pressure on the pistol to keep the hammer in place and push downward on the pistol in one continuous motion to cock the hammer. See figure 1-23.

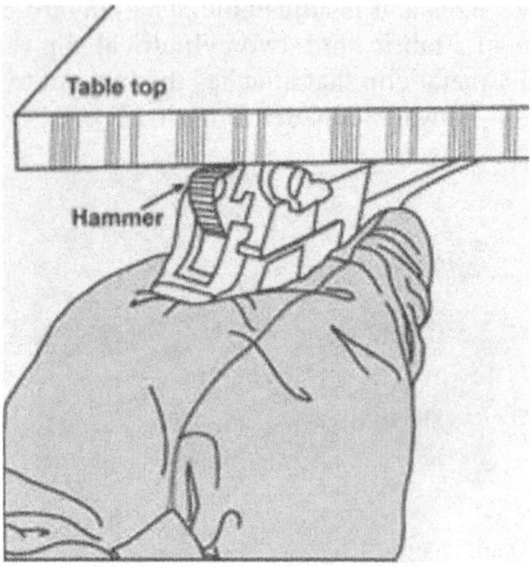

Figure 1-23. Cocking the Pistol on a Secure Surface.

Preventive Maintenance

If the M9 service pistol is to be effective, it must be maintained in a state of operational readiness at all times; therefore, maintenance of the M9 service pistol is a continuous effort. A clean, properly lubricated, well-maintained pistol will fire when needed.

Pistol Disassembly

Before disassembling the M9 service pistol, ensure that the pistol is in Condition 4. The pistol is in Condition 4 when the magazine is removed, the chamber is empty, the slide is forward, and the safety is on. To disassemble the pistol, perform the following steps in sequence:

- Hold the pistol in the right hand with the muzzle slightly elevated. Reach over the slide with the left hand and place the left index finger on the disassembly button and the left thumb on the disassembly lever. Press the disassembly button and hold it in place while rotating the disassembly lever downward until it stops.

Note: A left-handed Marine places the right thumb on the disassembly button and the right index finger on the disassembly lever.

- Pull the slide and barrel assembly forward and remove it while wrapping the fingers around the slide to hold the recoil spring and recoil spring guide in place.
- Turn the slide assembly over in the left hand until the recoil spring and recoil spring guide face up. Place the right thumb on the end of the recoil spring guide next to the locking block and compress the recoil spring and spring guide while lifting and removing them from the slide and barrel assembly. See figure 1-24. Allow the recoil spring to decompress slowly.

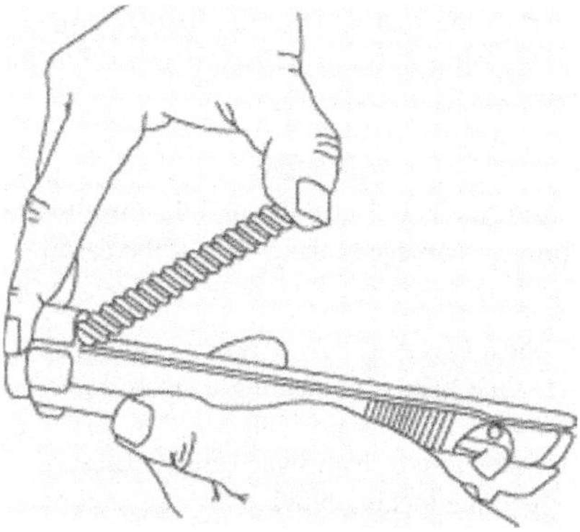

Figure 1-24. Removing the Recoil Spring and Recoil Spring Guide.

- Separate the recoil spring from the recoil spring guide.
- Push in on the locking block plunger with the right index finger while pushing the barrel forward slightly. Lift and remove the locking block and barrel assembly from the slide.

A Marine is not authorized to disassemble the pistol any further than the preceding steps. Any further disassembly must be performed by ordnance personnel. See figure 1-25.

Figure 1-25. Disassembled M9 Service Pistol.

Disassembly of the Magazine

To disassembly the magazine, perform the following steps:

- Grip the magazine firmly in the left hand with the floorplate up and the thumb resting against the flat end of the floorplate.
- Release the floorplate by pushing down (with a blunt object; e.g., an ink pen) on the floorplate retainer stud in the center of the floorplate. At the same time, slide the floorplate a short distance forward with the thumb.
- Maintain the magazine spring pressure with the thumb and remove the floorplate from the magazine.
- Remove the floorplate retainer and magazine spring and follower from the magazine tube.

See figure 1-26.

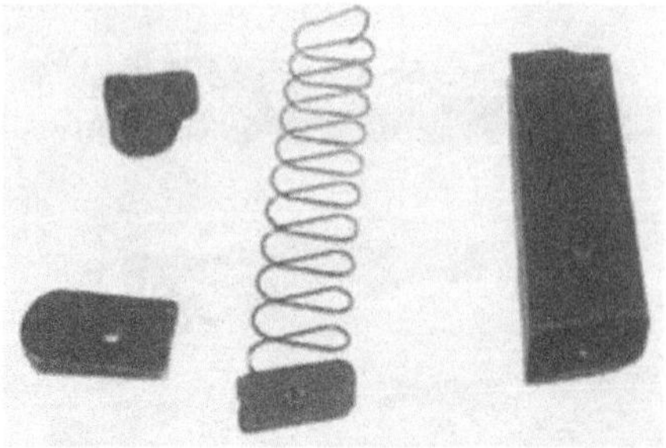

Figure 1-26. Disassembled Magazine.

Inspection of the Pistol

Once the pistol has been disassembled, it must be thoroughly inspected to ensure it is in a serviceable condition. Pistol inspection is continuous during the pistol's cleaning and reassembly:

Slide Assembly

- Check for free movement of the safety. Ensure the rear sight is secure.

Barrel Assembly

- Inspect the bore and chamber for pitting or obstructions.
- Check the locking block plunger for free movement of the locking block.
- Inspect the locking lugs for cracks and burrs.

Recoil Spring and Recoil Spring Guide

- Check the recoil spring for damage.
- Check that it is not bent.
- Check the recoil spring guide for straightness and smoothness.
- Check to be sure it is free of cracks and burrs.

Receiver Assembly

- Check for bends, chips, and cracks.
- Check for free movement of the slide stop and magazine catch assembly.
- Check the guide rails for excessive wear, burrs, cracks, or chips.

Magazine Assembly

- Check the spring and follower for damage.
- Ensure the lips of the magazine are not excessively bent and are free of cracks and burrs. The magazine tube should not be bent or dirty.

Cleaning and Lubricating the Pistol

Only authorized cleaning materials should be used to clean and lubricate the pistol. If these items are not issued with the pistol, they may be obtained from the armory. The following procedures are used to clean and lubricate the pistol—

Slide Assembly

- Clean the slide assembly with a cloth. A general purpose brush and cleaning lubricant protectant (CLP) can also assist in the removal of excess dirt and carbon buildup.
- Ensure the safety, bolt face, slide guides, and extractor are free of dirt and residue.
- Wipe dry with a cloth and apply a light coat of CLP.

Barrel Assembly

- Insert a bore brush with CLP into the chamber end of the barrel, ensuring that it completely clears the muzzle before it is pulled back through the bore.

Caution

Insert the bore brush through the chamber to prevent damage to the crown of the barrel.

- Repeat several times to loosen carbon deposits.
- Dry the barrel by pushing a swab through the bore.
- Repeat until a clean swab can be observed.
- Clean the locking block with a general purpose brush.
- Use the barrel brush to apply a light coat of CLP to the bore and chamber area and lubricate the exterior surfaces of the barrel and locking block.

Recoil Spring and Recoil Spring Guide

- Clean the recoil spring and recoil spring guide using CLP and a general purpose brush or cloth.
- Apply a light coat of CLP after wiping the recoil spring and recoil spring guide clean.

Receiver

- Wipe the receiver assembly clean with a cloth.
- Use a general purpose brush for areas that are hard to reach, paying special attention to the

disassembly lever, trigger, slide stop, hammer, and magazine release button.

- Apply a light coat of CLP.

Caution

Do not allow the hammer to fall with full force by pulling the trigger when the slide is removed. This can damage the receiver and hammer.

<u>Magazine</u>

- Clean the magazine tube and follower with CLP and a general purpose brush.
- Wipe the magazine spring, floorplate retainer, and floorplate clean with a cloth.
- Apply a light coat of CLP.

Pistol Re-assembly

After the M9 service pistol has been cleaned and lubricated, it must be properly re-assembled to ensure its serviceability. To re-assemble the M9 service pistol—

- Use the left hand to grasp the slide with the bottom facing up and the muzzle pointing toward the body. Use the right hand to grasp the barrel assembly with the locking block facing up. Use the index finger to push in the locking block plunger while placing the thumb on the base of the locking block.
- Insert the muzzle of the barrel assembly into the forward open end of the slide. At the same time, lower the rear of the barrel assembly by slightly moving the barrel downward. The locking block should fall into the notches of the slide assembly.
- Slip the recoil spring guide into the recoil spring.
- Insert the end of the recoil spring and recoil spring guide into the slide recoil spring housing. At the same time, compress the recoil spring and lower the spring guide until it is fully seated onto the locking block cutaway.
- Use the left hand to grasp the slide and barrel assembly, sights up, and wrap the fingers around the slide assembly to hold the recoil spring and guide in place. Align the slide assembly guide rails onto the receiver assembly guide rails.
- Push the slide rearward while pushing up on the slide stop with the thumb. Lock the slide to the rear while maintaining upward pressure on the slide stop. Rotate the disassembly lever upward. Listen for a click, an audible click indicates a positive lock.

Pistol Magazine Re-assembly

To re-assemble the magazine—

- Grip the magazine firmly in the left hand with the floorplate end up and the counting holes facing the Marine. Insert the follower into the magazine so the flat end of the follower is against the flat end of the magazine.
- Ensure the floorplate retainer is attached to the first curve of the bottom coil.
- Hold the spring upright with the right hand and insert the spring into the magazine tube so that the flat end of the floorplate retainer is against the flat end of the magazine.
- Push the magazine spring and floorplate retainer down with the right hand and hold it in place with the thumb of the left hand. Use the right hand to slide the floorplate over the side walls of the magazine until fully seated, which is indicated by an audible click.

Safety/Function Check

A safety/function check is performed after reassembling the M9 service pistol. Perform the following steps to ensure the pistol is operational:

- Ensure there is no ammunition in the chamber of the pistol.
- Ensure that the safety is in the safe position, then depress the slide stop, allowing the slide to return fully forward. At the same time, the hammer should fall to the full forward position.

- Pull and release the trigger. The firing pin block should move up and down but the hammer should not move.
- Place the safety in the fire position.
- Pull the trigger to check the double action. The hammer should cock and fall.
- Pull the trigger again and hold it to the rear. Use the fingers and thumb of the left hand to grasp the serrated sides of the slide just forward of the safety. Pull the slide to its rearmost position and release it while holding the trigger to the rear. Release the trigger, a click should be heard and the hammer should not fall.
- Pull the trigger to check the single action. The hammer should fall. Place the safety in the safe position.

If the safety/function check does not indicate an operational pistol, the Marine takes the pistol to organizational maintenance or the next authorized repair level.

User Serviceability Inspection

The Marine is responsible for performing a user serviceability inspection on the pistol prior to live fire. The user serviceability inspection ensures the pistol is in an acceptable operating condition. This inspection is not intended to replace the detailed pistol components inspection following disassembly or the limited technical inspection or pre-fire inspection conducted by a qualified armorer. To conduct a user serviceability inspection on the pistol, perform the following steps:

- Ensure the magazine release button is on the left side of the pistol for a right-handed Marine, the right side of the pistol for a left-handed Marine.
- Ensure the magazine seats into the magazine well when it is inserted and that it cannot be pulled out.
- Ensure the slide stays locked to the rear when the slide is pulled rearward with an empty magazine in the pistol.
- Ensure the magazine falls out freely when the magazine release button is depressed.
- Repeat the preceding four steps with the second magazine.
- Ensure the slide is locked to the rear before lubricating the spring guide, the top of the barrel just forward of the front sight, and the guide rails of the slide assembly behind the safety. With the muzzle pointed downward, work the slide several times and release.
- Inspect the pistol's external parts visually to ensure that there are no cracks or excessive wear.
- Perform a safety/function check of the pistol.

Pistol Maintenance in Adverse Conditions

Combat situations can place Marines in a variety of adverse conditions. Therefore, the M9 service pistol must be maintained properly to ensure its continued operation.

Extreme Cold

In extreme cold conditions, ensure that the following maintenance is performed:

- Clean and lubricate the pistol inside at room temperature, if possible.
- Apply a light coat of lubricant, arctic weather (LAW) to all functional parts.
- Always keep the pistol dry.
- Keep the pistol covered when moving from a warm to a cold area. This permit gradual cooling of the pistol and prevents freezing.
- Ensure that a hot pistol is not placed in snow or on ice.
- Keep snow out of the bore of the barrel. If snow should enter the bore, use a swab and cleaning rod to clean the bore before firing.

Hot, Wet Climates

Maintenance in hot, wet climates must be performed more frequently:

- Inspect hidden surfaces for corrosion. If corrosion is found, clean and lubricate.
- Remove handprints with a clean cloth in order to prevent corrosion.
- Dry the pistol with a cloth and lubricate it with CLP.
- Check ammunition and magazines frequently for corrosion. Disassemble and clean the magazines with CLP and wipe dry with a clean cloth. If necessary, clean ammunition with a dry cloth.

Hot, Dry Climates

In a hot, dry climate, dust and sand can cause stoppages and excessive wear on component contact surfaces during firing; therefore, keep the pistol covered whenever possible.

Corrosion is less likely to form on metal parts in a dry climate. Therefore, lightly lubricate internal working surfaces with CLP. Do not lubricate external parts of the pistol. Wipe excess lubricant from exposed surfaces. Do not lubricate internal components of the magazine.

Heavy Rain and Fording Operations

The following maintenance procedures are followed during periods of heavy rains or during fording operations:

- Always attempt to keep the pistol dry.
- Drain any water from the barrel prior to firing.
- Dry the bore with a swab and cleaning rod.
- Generously lubricate internal and external surfaces of the pistol with CLP.

Amphibious Conditions

If the pistol comes into contact with salt water, clean the pistol as soon as possible. If time does not permit cleaning in accordance with Technical Manual (TM) 1005A-10/1, *Operator's Manual, Pistol Semiautomatic, 9mm, M9*, then wash the pistol with fresh water

CHAPTER 2
WEAPONS HANDLING

Weapons handling is a method of providing consistent and standardized procedures for handling, operating, and employing the M9 service pistol. Understanding and applying the principles of weapons handling are critical to developing safe and consistent weapons skills. Strict adherence to training and diligent practice will make weapons handling instinctive. Mission accomplishment and survival during combat depend on a Marine's ability to react instinctively and with confidence.

Note: The procedures in this chapter are written for right-handed Marines; left-handed Marines should reverse directions as needed.

Safety Rules

The following safety rules are the foundation for responsible weapons handling. They must be observed at all times, both in training and combat.

Safety Rules

Rule 1: Treat every weapon as if it were loaded.
This rule is intended to prevent unintentional injury to personnel or damage to property from handling or transferring possession of a weapon.

Rule 2: Never point a weapon at anything you do not intend to shoot.
This rule enforces the importance of muzzle awareness and reinforces positive identification of the target.

Rule 3: Keep your finger straight and off the trigger until you are ready to fire.
This rule is intended to minimize the risk of firing the weapon negligently (when not firing, the trigger finger is straight along the receiver, outside of the trigger guard). This rule also reinforces positive identification of the target.

Rule 4: Keep weapon on safe until you intend to fire.
This rule enforces the use of the weapon's own safety feature and reinforces positive identification of the target.

Weapons Conditions

The M9 service pistol's level of readiness is defined by three specific conditions. The steps in the loading and unloading process take the pistol through the specific conditions that indicate the pistol's readiness for live fire. The Marine must understand and know the condition of his pistol at all times.

Condition 1. Magazine inserted, round in chamber, slide forward, and safety on.

Condition 2. Not applicable to the M9 service pistol.

Condition 3. Magazine inserted, chamber empty, slide forward, and safety on.

Condition 4. Magazine removed, chamber empty, slide forward, and safety on.

Determining a Weapon's Condition

There are two methods for determining the pistol's condition: checking the round indicator on the right side of the pistol and conducting a chamber check.

Checking the Round Indicator

When there is a round in the chamber, the upper surface of the extractor protrudes from the right side of the slide. The protrusion can be felt by sliding either the thumb or the index finger of the left hand over the top of the slide and across the extractor. See figure 2-1.

Conducting a Chamber Check

To conduct a chamber check—

- Point the pistol in a safe direction and grasp the pistol grip with the right hand.

Figure 2-1. Checking the Round Indicator.

- Place the trigger guard in the palm of the left hand. Use the thumb and index finger to grasp the forward end of the slide at the indentations under the front sight, behind the muzzle. See figure 2-2.

Figure 2-2. Placement of the Left Hand.

Caution

Ensure the muzzle does not cover the hand or fingers.

- Keep thumb in place around the pistol's backstrap and rotate fingers of the right hand over the top of the slide in front of the rear sight.
- Pull the slide to the rear by pushing forward with the right thumb and pulling back on the rear sight with the fingers. Use the left hand to steady the pistol and to assist in pulling the slide to the rear.
- Use the right hand to hold the slide to the rear (just enough to visually inspect the chamber for a round). Physically check for a round by inserting a finger of the right hand into the chamber area. See figure 2-3.

Figure 2-3. Chamber Check.

Note: At night or in low light conditions, the Marine's visibility is reduced; therefore, the Marine will have to rely on the physical check with the finger to determine if a round is in the chamber.

Caution

Pulling the slide too far to the rear while inspecting the chamber may cause a double feed or the ejection of a round.

- Remove the finger from the chamber and release tension on both hands to allow the slide to go forward. Ensure that the slide is all the way forward.

Weapons Commands

Weapons commands direct the Marine to safely load, unload, and employ the M9 service pistol. Six commands are used in weapons handling:

"*Load*"
This command is used to take a weapon from Condition 4 to Condition 3.

"*Make Ready*"
This command is used to take a weapon from Condition 3 to Condition 1.

"*Fire*"
This command is used to engage targets.

"*Cease Fire*"
This command is used to cease target engagement.

"*Unload*"
This command is used to take a weapon from any condition to Condition 4.

"*Unload, Show Clear*"
This command is used to require a second individual to check the weapon to verify that no ammunition is present before the weapon is put into Condition 4.

Figure 2-4. Indexing the Magazine.

Loading the Pistol

Perform the following steps to load the pistol (take the pistol from Condition 4 to Condition 3):

- Ensure the pistol is on safe.
- Use the right hand to grip the pistol grip firmly. Ensure that the pistol is pointed in a safe direction, bring the trigger guard to the right of eye level and cant the pistol so the magazine well faces inboard at approximately a 45-degree angle to the deck. Draw the right elbow in to facilitate control of the pistol.
- Use the left hand to remove a filled magazine from the ammunition pocket. Index the magazine by sliding the index finger along the forward edge of the magazine. See figure 2-4.
- Insert the filled magazine into the magazine well by guiding it with the index finger and, with the fingers extended, pushing it in with the heel of the hand until it is fully seated. Do not relinquish contact with the magazine until it is fully seated. See figure 2-5.

Figure 2-5. Seating the Magazine.

Making the Pistol Ready

Perform the following steps to take the pistol from Condition 3 to Condition 1:

- Firmly grip the pistol grip with the right hand. Ensure that the pistol is pointed in a safe direction and the slide is in its forward position.
- Rotate the magazine well outboard to facilitate pulling the slide to the rear. With the fingers and thumb of the left hand, grasp the serrated sides of the slide just forward of the safety. See figure 2-6.

Caution

Ensure the muzzle does not cover the hand or fingers.

- Pull the slide to its rearmost position by pushing forward with the right hand while pulling back on the slide with the left hand.
- Release the slide, this strips a round from the magazine and chambers it as the slide moves forward.
- Ensure the pistol remains on safe.
- Conduct a chamber check to ensure a round is in the chamber.

Note: A chamber check may be conducted at any time to check the pistol's condition.

Fire

Perform the following steps to fire the pistol:

- Keep trigger finger straight and use the right thumb to take the pistol off safe.
- Place the trigger finger on the trigger and apply pressure to the trigger until the shot is fired.

Right Side View

Left Side View

Figure 2-6. Grasping the Slide to Make Ready.

Cease Fire

Perform the following steps to execute a cease fire of the pistol:

- Remove the finger from the trigger and place it straight along the receiver.
- Place the pistol on safe without breaking the grip of the right hand.
- Assume a carry or transport position.

Unloading the Pistol

Perform the following steps to take the pistol from any condition to Condition 4:

- Use the right hand to grip the pistol firmly. Ensure that the pistol is on safe.
- Rotate the pistol so the magazine well is pointed inboard and angled down.

  Note: The angle of the magazine well must allow the magazine to fall freely from the well once the magazine release button is engaged.

- Depress the magazine release button to remove the magazine from the pistol. Catch the magazine with the left hand and retain it.
- Push upward on the slide stop with the right thumb and maintain pressure. Rotate the weapon so the chamber is outboard.

  Note: A left-handed Marine pushes upward on the slide stop with the left index finger.

- Reach over the top of the pistol with the left hand and grasp the slide serrations with the thumb and index finger. The left hand should partially cover the ejection port so it is positioned to catch an ejected round.
- Point the pistol in a safe direction and fully retract the slide and lock it to the rear. At the same time, catch the ejected round with the left hand. See figure 2-7.

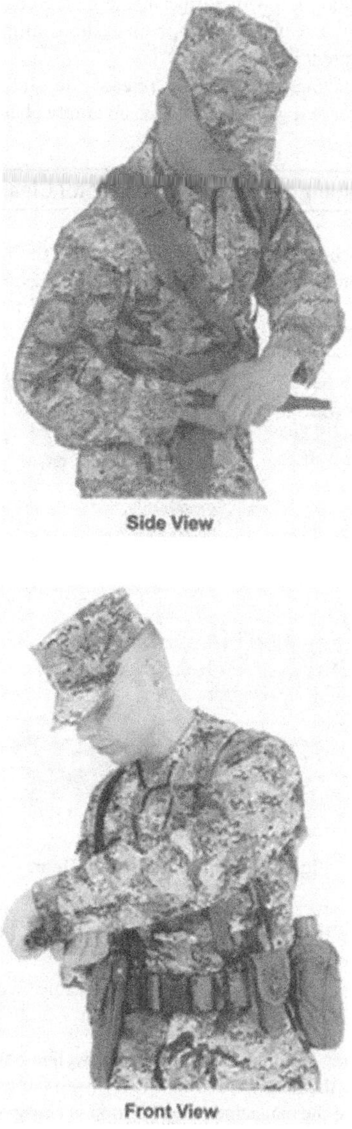

Figure 2-7. Catching the Ejected Round.

- Rotate the pistol so the inside of the chamber can be seen. Visually inspect the chamber to ensure it is empty.
- Press the slide stop to release the slide and observe it going forward on an empty chamber.

Unloading and Showing the Pistol Clear

Perform the following steps to take the pistol from any condition to Condition 4. See figure 2-8.

Figure 2-8. Unload, Show Clear.

- Use the right hand to grip the pistol firmly. Ensure that the pistol is on safe.
- Rotate the pistol so the magazine well is pointed inboard and angled down.

  Note: The angle of the magazine well must allow the magazine to fall freely from the well once the magazine release button is engaged.

- Depress the magazine release button to remove the magazine from the pistol. Catch the magazine with the left hand and retain it.
- Push upward on the slide stop with the right thumb and maintain pressure. Rotate the weapon so the chamber is outboard.

  Note: A left-handed Marine pushes upward on the slide stop with the left index finger.

- Reach over the top of the pistol with the left hand and grasp the slide serrations with the thumb and index finger. The left hand should partially cover the ejection port so it is positioned to catch an ejected round.
- Point the pistol in a safe direction and fully retract slide and lock it to the rear. At the same time, catch the ejected round with the left hand.
- Rotate the pistol so the inside of the chamber can be seen. Visually inspect the chamber to ensure it is empty.
- Bring the pistol to the administrative transport and have another Marine visually inspect the chamber to ensure that—
  - The chamber is empty, no ammunition is present, and the magazine is removed.
  - The pistol is on safe.
- Acknowledge that the pistol is clear.
- Press the slide stop to release the slide and observe it going forward on an empty chamber.

Emptying the Magazine

Once the pistol is unloaded, the pistol magazine can be emptied of ammunition. To empty the magazine, perform the following steps:

- Hold the magazine upright with the back of the magazine tube against the palm of the hand.
- Push the top round forward with the thumb and catch it with the other hand as it is removed.
- Repeat until the magazine is empty.

Filling the Magazine

Prior to loading the pistol, the pistol magazine must be filled with the prescribed number of rounds of ammunition. See figure 2-9. To fill the magazine, perform the following steps:

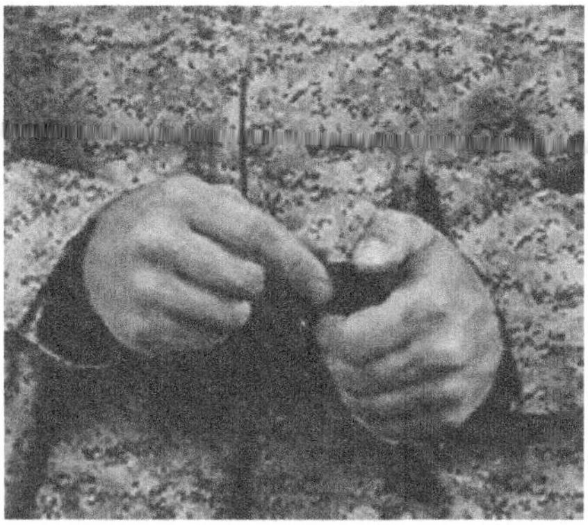

Figure 2-9. Filling the Magazine.

- Hold the magazine with the back of the magazine against the palm of the hand and the follower up.
- Use the other hand to place a round (primer end first) on the follower in front of the magazine lips.
- Press down on the round and slide the round completely back under the lips. The thumb or finger may push down on the back of the round to assist movement. The base of the round should be flush with the back of the magazine.

Repeat this procedure until the magazine is filled with the appropriate number of rounds. Holes on the back of the magazine allow the visual counting of rounds in five round increments.

Reloading the Pistol

The Marine's ability to reload the pistol quickly improves his chance for success on the battlefield.

Dry Reload

A dry reload is conducted when the pistol runs out of ammunition during engagement and the slide locks to the rear. See figure 2-10. Perform the following steps to conduct a dry reload with the slide locked to the rear:

Note: The pistol is not placed on safe during a dry reload; the trigger finger is taken out of the trigger guard and placed straight along the side of the receiver.

- Seek cover, if the situation permits.

Figure 2-10. Pistol Ran Dry of Ammunition.

- Retain the firing grip with the right hand and pull the pistol in close to the body to facilitate control. Bring the trigger guard to the right of eye level and cant the pistol so the magazine well is facing inboard at approximately a 45-degree angle to the deck. See figure 2-11.

Figure 2-11. Bring the Pistol Toward the Body.

- Press the magazine release button and let the empty magazine fall to the deck. At the same time, unfasten the ammunition pocket to withdraw a filled magazine. See figure 2-12.

  Note: The primary objective during a dry reload is to get the pistol back in action as quickly as possible. Following engagement, retrieve the magazine before moving.

- Grasp the magazine by curling the middle finger and thumb of the left hand around the base of the magazine, with the index finger straight along the ammunition pocket. See figure 2-13.

Figure 2-12. Release Magazine and Unfasten Ammunition Pocket.

Figure 2-13. Grasping the Magazine.

- Index the magazine: as the magazine is being withdrawn from the pocket, the index finger should be along the front of the magazine. See figure 2-14.

Figure 2-14. Indexing the Magazine.

- Rotate the hand up so the magazine is aligned with the magazine well.
- Glance quickly at the magazine well, insert the magazine into the magazine well (see fig. 2-15).

Figure 2-15. Glance at the Magazine Well.

- Bring the eyes back on target at the same time as the heel of the left hand seats the magazine, do not relinquish contact with the magazine. See figure 2-16.

Figure 2-16. Seating the Magazine.

- Roll both hands inward to establish a two-handed grip and press the slide release with the left thumb to allow the slide to move forward, chambering the first round, and present the pistol to the target. See figure 2-17.

Note: A left-handed Marine presses the slide release with his trigger finger.

Figure 2-17. Present Pistol Back to Target.

Figure 2-18. Facilitating Control of Pistol.

Condition 1 Reload

In a Condition 1 reload, a partially-filled magazine is removed from the pistol and replaced with a fully filled magazine. A Condition 1 reload is performed when there is a lull in the action or whenever deemed necessary by the Marine. To perform a Condition 1 reload—

- Retain the firing grip with the right hand and pull the pistol in close to the body to facilitate control. Keep the pistol pointed in the direction of the likely threat. See figure 2-18.
- Withdraw a filled magazine from the ammunition pocket with the left hand. Index the magazine and bring it up to the left of eye level. See figure 2-19.

Figure 2-19. Withdrawing and Indexing a Filled Magazine.

- Slide the index finger to the side of the magazine to grasp the magazine between the index and middle fingers. See figure 2-20.

Figure 2-20. Grasping a Filled Magazine.

- Raise the pistol and bring the trigger guard to the right of eye level and cant the pistol so that the magazine well is facing inboard at approximately a 45-degree angle to the deck.

  Note: The angle of the magazine well must allow the magazine to fall freely from the well once the magazine release button is engaged.

- Press the magazine release button with the right thumb to eject the partially-filled magazine from the magazine well. Grasp the magazine between the index finger and thumb. See figure 2-21.

Figure 2-21. Removing a Partially-Filled Magazine.

- Insert the filled magazine into the magazine well (see fig. 2-22). Use the heel of the hand to ensure it is fully seated (see fig. 2-23).

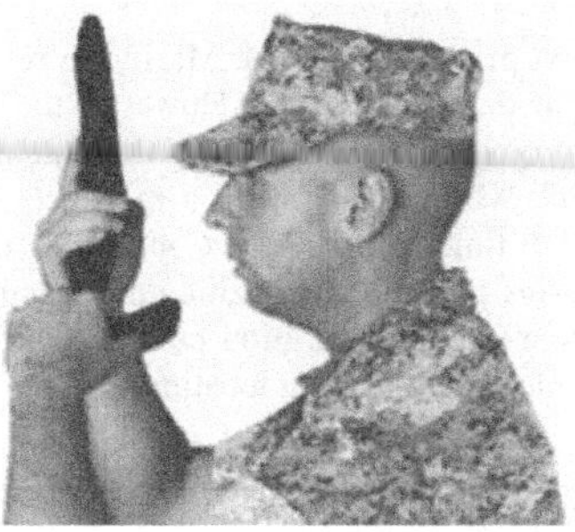

Figure 2-22. Inserting a Filled Magazine.

Figure 2-23. Seating a Filled Magazine.

- Lower the pistol and point it in the direction of the likely threat.

If time permits, examine the partially-filled magazine to determine the number of rounds remaining. Stow the partially-filled magazine in the ammunition pocket for later use.

Reloading Considerations

If possible, take cover before reloading. Always reload before leaving cover to take advantage of the protection.

When reloading, the first priority is to reload the pistol quickly so that it is ready to fire. During a reload, the Marine focuses on reloading only—not on the enemy.

The next priority is for the Marine to retain the magazine during the reload. However, the combat situation may dictate dropping the magazine to the deck when performing a reload (i.e., dry reload). If time permits (i.e., Condition 1 reload), the Marine picks the magazine up or secures the magazine (e.g., ammunition pocket, flak jacket) before moving to another location.

Remedial Action

The M9 service pistol is an effective and extremely reliable weapon. Proper care and preventive maintenance usually ensures the pistol's serviceability. However, stoppages, while infrequent, do occur. To keep the pistol in action, stoppages must be cleared as quickly as possible through remedial action. A malfunction cannot be corrected through remedial action by the Marine.

Stoppage

A stoppage is an unintentional interruption in the cycle of operation; e.g., the slide not moving forward completely. A stoppage is normally discovered when the pistol will not fire. Most stoppages can be prevented by proper care, cleaning, and lubrication of the pistol.

Many stoppages of the M9 service pistol are caused by shooter error. The Marine must be aware of shooter-induced stoppages in order to avoid them or to quickly identify and correct the stoppage and return the pistol to action. In a shooter-induced stoppage, the Marine—

- Fails to make ready.
- Fails to take the pistol off safe prior to firing.
- Engages safety while firing.
- Engages magazine release button while firing.
- Engages slide stop while firing (particularly prevalent when firing with an Isosceles grip).
- Fails to reset the trigger.
- Fails to recognize the pistol has run dry and the slide has locked to the rear.

Malfunction

A malfunction is a failure of the pistol to fire satisfactorily or to perform as designed (e.g., a broken front sight that does not affect the functioning of the pistol). A malfunction does not necessarily cause an interruption in the cycle of operation. When a malfunction occurs, the pistol must be repaired by an armorer.

Remedial Action

There is no one set of procedures (i.e., immediate action) that can be performed to clear all or even most of the stoppages that can occur with the M9 service pistol. Therefore, remedial action requires investigating the cause of the stoppage, clearing the stoppage, and returning the pistol to operation. When performing remedial action, the Marine should seek cover if the tactical situation permits. Once a pistol ceases to fire, the Marine must visually or physically observe the pistol to identify the problem before it can be cleared:

> *Note:* The steps taken to clear the pistol are based on what is observed.

- Remove the finger from the trigger and place it straight along the receiver.
- Bring the pistol in close to the body and in a position to observe the chamber.
- Pull the slide to the rear while observing the chamber area to identify the stoppage. See figure 2-24 on page 2-14.

> *Note:* Ensure the pistol does not move to safe when pulling the slide to the rear.

Figure 2-24. Observing Chamber.

- Correct the stoppage:
  - If there is a round in the magazine but not in the chamber (see fig. 2-25), the slide is released and a round is observed being chambered.

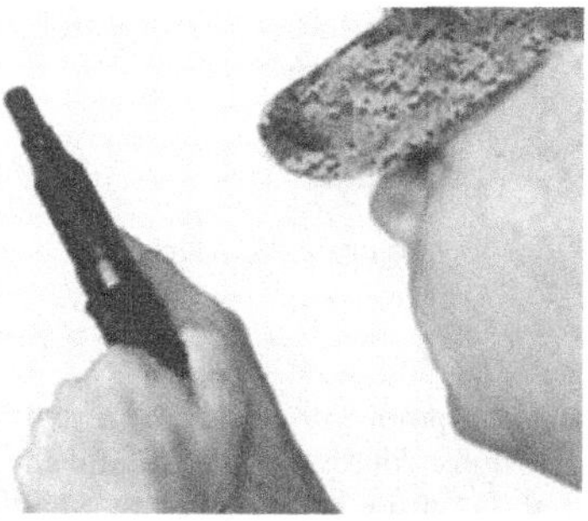

Figure 2-25. Round in Magazine but Not in Chamber.

  - If a round being chambered is not observed, the bottom of the magazine is tapped to seat it properly, and the slide is racked to the rear. See figure 2-26.
  - If there is no round in the magazine or chamber, a reload is conducted. See figure 2-27.

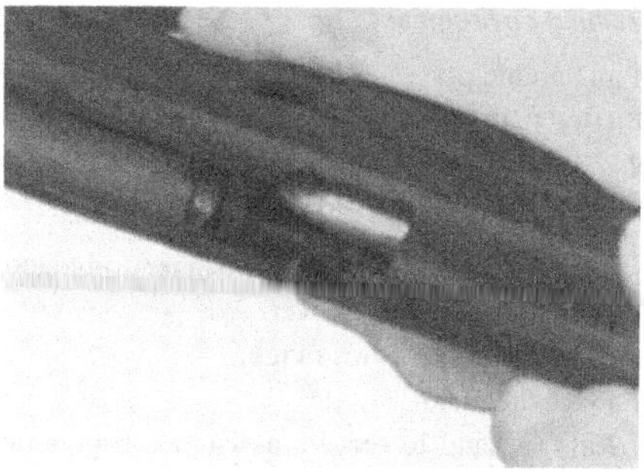

Figure 2-26. Round Not Being Chambered.

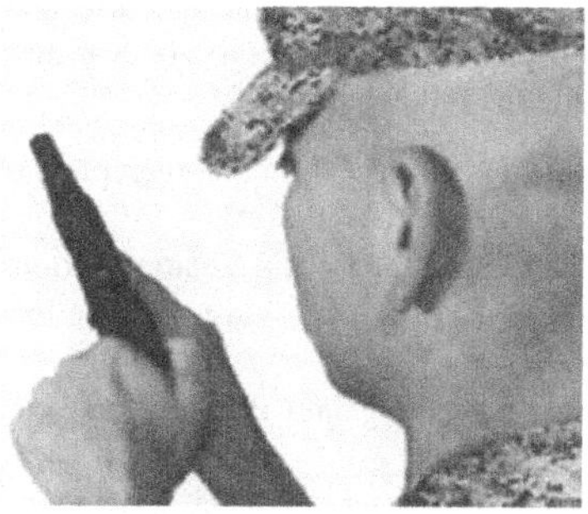

Figure 2-27. No Round in Magazine or Chamber.

- Fire the pistol.

Audible Pop or Reduced Recoil

WARNING

When an audible pop or reduced recoil is experienced, the Marine DOES NOT perform remedial action unless he is in a combat environment. An audible pop occurs when only a portion of the propellant is ignited. It is normally identifiable by reduced recoil and the pistol will not cycle. Sometimes, it is accompanied by excessive smoke escaping from the chamber area.

Training Environment

If an audible pop or reduced recoil is experienced during firing, cease fire immediately. Do not apply remedial action; instead, perform the following steps:

- Remove the finger from the trigger and place it straight along the receiver.
- Point the pistol down range.
- Place the pistol on safe.
- Raise a hand to receive assistance from available range personnel.

Combat Environment

The tactical situation may dictate correction of an audible pop or reduced recoil. To clear the pistol, perform the following steps:

- Remove the finger from the trigger and place it straight along the receiver.
- Seek cover if the tactical situation permits.
- Unload the pistol, but leave the slide locked to the rear.
- Insert something into the bore and clear the obstruction.
- Observe the barrel for cracks or bulges.
- Reload the pistol.

Weapons Carries

As the threat level increases, so should the Marine's readiness for engagement. Weapons carries are designed to place the Marine in a state of increased readiness as the threat level increases. There are two carries with the pistol: the Alert and the Ready. The carries permit quick engagement when necessary.

Alert

The Alert is used when enemy contact is likely (probable). See figure 2-28. The Marine performs the following steps to assume the Alert:

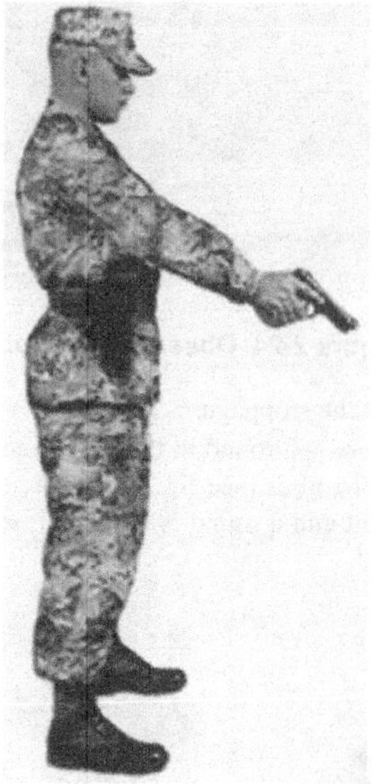

Figure 2-28. Alert.

- Ensure the pistol is on safe.
- Grasp the pistol grip firmly with two hands. The trigger finger is straight and the right thumb is on the safety and in a position to

- Extend the arms down at approximately a 45-degree angle to the body or bend the elbows. See figure 2-29.
- The muzzle of the pistol is pointed in the likely direction of the threat.

Ready

The Ready is used when there is no target, but contact with the enemy is imminent. The Marine performs the following steps to assume the Ready (see fig. 2-30):

- Ensure the pistol is on safe.
- Grasp the pistol firmly with two hands. The trigger finger is straight and the right thumb is on the safety and in a position to operate it.
- Extend arms and raise the pistol to just below eye level so a clear field of view is maintained.
- Point the muzzle of the pistol in the direction of enemy contact.

Weapons Transports

The M9 service pistol is transported in either the holster transport or the administrative transport.

Holster Transport

The holster transport is the most common method of carrying the pistol because it can be transported safely in the holster. This transport is used when there is no immediate threat (enemy contact is remote). See figure 2-31. To transport the pistol in the holster:

- Point the pistol in a safe direction.

WARNING

Ensure the pistol is pointed in a safe direction at all times and does not cover any part of the body while holstering.

Figure 2-29. Alert—Close Quarters.

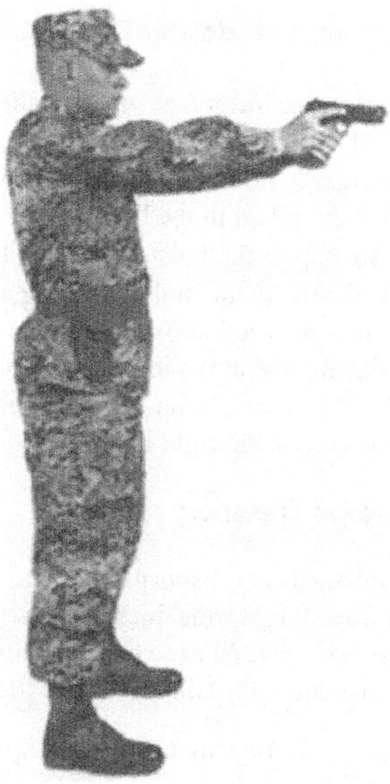

Figure 2-30. Ready.

Figure 2-31. Holster Transport.

Figure 2-32. Administrative Transport.

- Ensure that the safety is on, the slide is forward, and the trigger finger is straight.
- Use the right hand to firmly grip the pistol grip and place the pistol in the holster:
  - Lift the flap of the holster with the left hand.
  - Look down at the holster, bring the pistol back to a position above the holster, and rotate the muzzle down into the holster.
- Push the pistol snugly into the holster and fasten the flap with the right hand.

Administrative Transport

The administrative transport is used to transport the pistol when the Marine does not have a holster. See figure 2-32. The Marine performs the following steps to assume the administrative transport:

- Establish a firm grip around the pistol grip with the right hand.
- Ensure the pistol is on safe, the magazine is removed, the slide is locked to the rear, and the trigger finger is straight along the receiver.
- Bend the elbow to approximately a 45-degree angle so the pistol is positioned near shoulder level. The wrist should be straight so the pistol's muzzle points up.

Combat Mindset

In a combat environment, the Marine must be constantly prepared to engage targets. When a target presents itself, there may be little time to react. The target must be engaged quickly and accurately. It is not enough to simply know marksmanship techniques, the Marine must be able to react instinctively.

The development of a combat mindset can be associated with the carries and holster transport for the pistol. The use of each carry/transport is dictated by the perceived level of threat. Each carry and transport should signify a stage of mental and physical preparedness for combat. The intensity of the Marine's mental and physical preparation depends on the likelihood of enemy contact.

Physical and Mental Preparation

Physical Preparation

In combat, targets can present themselves without warning. Therefore, it is essential for the Marine to maintain proper balance and control of the pistol at all times so that the pistol can be presented quickly and the target engaged accurately.

Speed alone does not equate to effective target engagement. The Marine should fire only as fast as he can fire accurately, never exceeding his physical ability to apply the fundamentals of marksmanship. To be effective in combat, the Marine must train to perfect the physical skills of shooting so they become second nature. The more physical skills that can be performed automatically, the more concentration that can be given to the mental side of target engagement.

Mental Preparation

While combat is unpredictable and constantly changing, the Marine can prepare mentally for the contingencies of the operational setting and confrontation with a threat. The stress of combat, coupled with the limited time available to engage targets, requires concentration on the mental aspects of target engagement; e.g., identification of targets, shoot/no-shoot decisionmaking, and the selection and use of cover. Minimizing stress and maximizing the limited time available to engage targets can be accomplished by—

- Knowing the combat environment and being constantly aware of the surroundings (e.g., terrain, available cover, possible threats) enables the Marine to quickly present the pistol and accurately engage targets.
- Identifying and evaluating possible courses of action and developing potential plans for target engagement that will be appropriate to the combat situation.
- Instilling confidence in a Marine's ability to fire well-aimed shots in the stress of a combat situation. A key factor in a Marine's level of confidence is the degree to which he has mastered the tactics, techniques, and procedures of pistol marksmanship.

Threat Levels

No Immediate Threat

When there is no immediate threat, the Marine assumes the holster transport. The pistol should be in Condition 1. This is the lowest level of awareness for the Marine in a combat environment, but the Marine must stay alert and aware of any nearby activity. To prepare for target engagement at this level, the Marine must—

- Be aware of likely areas of enemy contact.
- Be aware of the condition of his pistol.
- Establish a plan or course of action to present the pistol to a target should a target appear.
- Mentally review appropriate actions such as reloading and remedial action.

Contact Likely (Probable)

If enemy contact is likely (probable), the Marine assumes the Alert. When enemy contact is likely, the Marine should—

- Expect enemy contact and be constantly prepared to present the pistol.
- Search the entire area for indications of enemy targets and for suitable terrain features that offer cover and concealment. The Marine should avoid restricting the search to a single terrain feature because this hinders awareness to a sector of the battlefield and to approach by the enemy.
- Be mentally prepared for contact. Plan a course of action for immediate response to a target. Modify the plan of action as needed.

- Be physically prepared to fire. Maintain proper balance at all times. Avoid self-induced physical fatigue. For example, do not grip the pistol so tightly that fingers, hands, and arms tire from carrying the pistol.

Contact Imminent

When contact with an enemy target is imminent, assume the Ready. In this carry, the Marine is at the highest level of awareness and is constantly searching for and expecting a target. To fire well-aimed shots upon target detection, the Marine must be at the peak of his mental preparation, all distractions must be eliminated, and his focus must be on firing an accurate shot. In the Ready, the Marine must—

- Keep the pistol oriented in the general direction of observation (eyes, muzzle, target).
- Maintain a clear field of view above the pistol sights until the target is detected.
- Be mentally and physically prepared to engage the target. The Marine must be ready to:
  - Identify the target.
  - Sweep the safety.
  - Apply the fundamentals of marksmanship.
- Move only as fast as he is capable of delivering well-aimed shots, ensuring that speed of engagement does not exceed his physical abilities.
- Search the entire area for indications of enemy targets, lowering the pistol enough to observe a clear field of view of the area.

Transferring the Pistol

The Marine's ability to transfer a pistol to another Marine is critical to safe weapons handling. There are two methods for transferring the pistol from one Marine to another: show clear transfer and condition unknown transfer. Each transfer is performed based on the operational environment/combat situation.

Show Clear Transfer

To transfer the pistol—

- Grasp the pistol firmly in the right hand while ensuring that the pistol is on safe.
- Remove and retain the magazine.
- Lock the slide to the rear and catch the round if there is a round in the chamber.
- Inspect the chamber visually to ensure it is empty and leave the slide locked to the rear. To transfer the pistol—
  - If the receiving Marine is to the right: Cradle the trigger guard in the palm of the left hand and wrap the fingers around the top of the pistol. Release the firing grip.
  - If the receiving Marine is to the left: With the left hand, grasp the slide of the pistol with the thumb over the slide and the fingers underneath. Release the firing grip.
- Ensure that the muzzle points up at a 45-degree angle in a safe direction and the chamber exposed. Hand the pistol to the other Marine, grip first. See figure 2-33.

Figure 2-33. Show Clear Transfer.

- The receiving Marine—
  - Grasps the pistol grip with the trigger finger straight along the receiver.
  - Inspects the chamber visually to ensure it is empty.
  - Ensures the pistol is on safe.

Condition Unknown Transfer

To transfer the pistol—

- Grasp the pistol firmly in the right hand while ensuring that the pistol is on safe. To transfer the pistol—
  - If the receiving Marine is to the right: Cradle the trigger guard in the palm of the left hand and wrap the fingers around the top of the pistol. Release the firing grip. See figure 2-34.

Figure 2-34. Condition Unknown Transfer to the Right.

  - If the receiving Marine is to the left: With the left hand, grasp the slide of the pistol with the thumb over the slide and the fingers underneath. Release the firing grip. See figure 2-35.

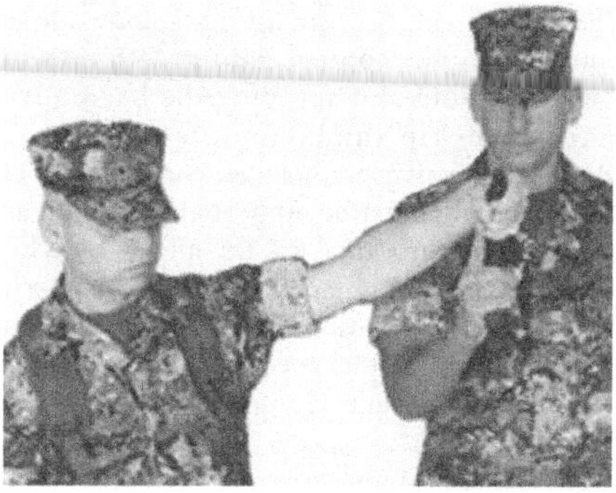

Figure 2-35. Condition Unknown Transfer to the Left.

- Ensure that the muzzle points up at a 45-degree angle in a safe direction. Hand the pistol to the receiving Marine, grip first.
- The receiving Marine—
  - Grasps the pistol grip with the trigger finger straight along the receiver.
  - Ensures the pistol is on safe.
  - Conducts a chamber check to determine the condition of the pistol.
  - Remove the magazine and count the number of rounds in the magazine by using the counting holes, if time permits. Re-insert the magazine into the magazine well ensuring it is fully seated.

CHAPTER 3
FUNDAMENTALS OF PISTOL MARKSMANSHIP

The fundamentals of pistol marksmanship are aiming, trigger control, and breath control. Understanding and applying the basic pistol marksmanship fundamentals ensures the Marine's effectiveness in target engagement. The fundamentals must be continually studied and practiced because they are the means by which accurate shots are placed on target. A Marine with a solid foundation in the fundamentals of marksmanship will be successful in the application of these fundamentals during combat.

Aiming

Maintaining the correct relationship between the pistol sights is essential for accurate target engagement. Because of the short distance between the pistol sights, a small error in their alignment causes a considerable error at the target.

Sight Alignment

Sight alignment is the relationship between the front sight and rear sight with respect to the aiming eye. Correct sight alignment is the front sight centered in the rear sight notch with the top edge of the front sight level **aligned** with the top edge of the rear sight. There should be equal space on either side of the front sight. See figure 3-1.

Figure 3-1. Sight Alignment.

Establishing Sight Alignment

The pistol is fired without benefit of bone support; therefore, the pistol is in constant motion. The Marine must understand this, yet continually strive to align the sights. To fire accurately, the sights must be aligned when the shot breaks.

Grip

The grip is key to acquiring sight alignment. If the grip is correct, the front and rear sights should align naturally. Dry fire during presentation of the M9 service pistol aids in obtaining a grip that allows sight alignment to be acquired consistently.

Controlled Muscular Tension

There must be enough controlled muscular tension in the grip, wrists, and forearms to hold the pistol steady and level the barrel to maintain sight alignment. Consistent tension stabilizes the sights and maintains sight alignment.

Sight Picture

Sight picture is the placement of the front sight in relation to the target while maintaining sight alignment. See figure 3-2 on page 3-2.

Because the pistol is constantly moving, sight picture is acquired within an aiming area that is located center mass on the target. The aiming area allows for movement of the sights on the target while maintaining sight alignment. Each Marine defines an acceptable aiming area within his own ability to stabilize the sights. Time, distance to the target, and personal ability affect dictate the aiming area. As the Marine becomes more proficient with the pistol, the aiming area becomes more precise.

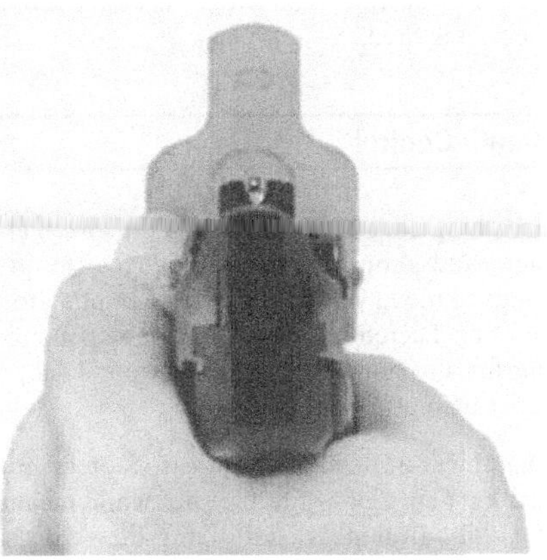

Figure 3-2. Sight Picture.

The aiming area is determined by the Marine's stability of hold. The proper grip stabilizes the sights so sight alignment can be maintained, but the sights move continuously within the aiming area of the target. The Marine understands the pistol's movement and learns to apply trigger control as he is obtaining sight alignment/sight picture within the aiming area so the shot breaks the moment sight picture is established.

Relationship Between the Eye and the Sights

The human eye can focus clearly on only one object at a time. The Marine must focus on the top edge of the front sight and fire the shot while maintaining the relationship between the front and rear sights within the aiming area. Focusing on the top edge of the front sight rather than the target keeps the front sight clear and distinct, which allows the Marine to detect minor variations in sight alignment. Secondary vision allows the Marine to see the target (although slightly blurred) and maintain sight picture within his aiming area.

Trigger Control

Trigger control is the Marine's skillful manipulation of the trigger that causes the pistol to fire while maintaining sight alignment and sight picture. Proper trigger control aids in maintaining sight alignment while the shot is fired.

Sight Alignment and Trigger Control

Aiming and trigger control are mutually supportive—one cannot be performed without the other. Sight alignment and trigger control must be performed simultaneously to fire an accurate shot. As pressure is applied to the trigger, the sights may move, causing them to be misaligned. To fire accurate shots, the sights must be aligned when the shot breaks. Trigger control can actually assist in aligning the sights. With proper trigger finger placement and consistent muscular tension applied to the grip, the sights can be controlled as the trigger is moved to the rear. If the sights move extensively while pressing the trigger, this can indicate an improper grip or inconsistency in the muscular tension being applied to the grip.

Grip

A firm grip is essential for good trigger control. The grip is established before applying trigger control and is maintained throughout the firing process. To establish the grip, the hand is placed around the pistol grip in a location that allows the trigger finger to move the trigger straight to the rear while maintaining sight alignment. Once the grip is established, it should be firm enough to allow manipulation of the trigger while maintaining sight alignment. The pressure applied to the grip must be equal to or more than the pressure required to move the trigger to the rear. If the pressure is not applied correctly, the sights move as the trigger is pressed to the rear and sight alignment is disturbed.

Trigger Finger Placement

Once the grip is established, the finger is placed on the trigger. Placement of the finger should be natural and allow free movement of the trigger finger. A natural trigger finger placement allows the trigger to be moved straight to the rear while maintaining sight alignment. If the finger presses the trigger to the side, it can cause an error in sight alignment and shot placement.

Each Marine must experiment with finger placement in order to determine effective placement on the trigger. Once established, effective trigger finger placement allows the trigger to be consistently moved straight to the rear while maintaining sight alignment.

Types of Trigger Control

Uninterrupted Trigger Control

During uninterrupted trigger control, the Marine applies a steady, unchanging pressure to the trigger until the shot is fired. Uninterrupted trigger control is particularly effective at close range, when the target area is large, and when stability of hold is not critical for accuracy. To apply uninterrupted trigger control, apply pressure on the trigger while maintaining focus on the top edge of the front sight. Continue pressure on the trigger to begin moving the trigger straight to the rear while obtaining sight alignment and sight picture. Move the trigger straight to the rear in a single, smooth motion with no hesitation.

Interrupted Trigger Control

Interrupted trigger control is particularly effective at longer ranges, when the target is small, and when stability of hold is critical to maintaining sight picture in the aiming area. This method is also used if the pistol sights move outside the aiming area when applying trigger control. If the sight picture is outside the aiming area, the Marine stops and holds the rearward movement on the trigger until sight picture is re-established. When sight picture is re-established, the rearward movement of the trigger is continued until the shot is fired.

Breath Control

Breathing causes movement of the chest, abdomen, and shoulders, which causes the pistol sights to move vertically while attempting to aim and fire. Therefore, it is necessary to stop breathing for a period of time while firing a shot or a series of shots.

The object of breath control is to stop breathing just long enough to fire the shot while maintaining sight alignment, stabilizing the sights, and establishing the sight picture. To be consistent, the breath should be held at the same point in the breathing cycle; i.e., the natural respiratory pause.

Breathing should not be stopped for too long because it has adverse visual and physical effects. Holding the breath longer than is comfortable results in a lack of oxygen that causes vision to deteriorate and then affects the ability to focus on the sights.

Application of Marksmanship Fundamentals in Field Firing

Compression of the Fundamentals

Pistol engagements typically occur over close distances and are short in duration. Because an immediate response to the threat is required, the application of the fundamentals must be a conditioned response that is executable in a compressed time. The goal of successful, quick target engagement is the application of the fundamentals of marksmanship while firing the shot the moment weapon presentation is complete.

The time required to move the trigger to the rear while acquiring and maintaining sight alignment and sight picture is unique to each Marine and is based on his capabilities. Each Marine should know his abilities and fire only as quickly as he is capable of firing accurately. The Marine must not exceed his shooting skills in an effort to get rounds off quickly.

Aiming

In field firing, the fundamentals are applied in a compressed time so sight alignment and sight picture are achieved as the shot is fired. Although the target must be quickly engaged in combat, sight alignment is still the first priority: strive for a clear front sight. Distance to the target and the size of the target affects sight alignment as follows:

- As the distance to the target increases and the size of the target decreases, sight alignment becomes more critical to target engagement. Accurate sight picture/sight alignment cannot be compromised for speed.
- Sight alignment is critical to the effective engagement of smaller targets such as partially exposed targets.
- As the distance to the target decreases, perfect sight alignment is not as critical, but there must be a relationship between the sights and their placement on the target within the aiming area to ensure accuracy.

Trigger Control

Proper trigger control aids in maintaining sight alignment while the shot is fired. As pressure is applied to the trigger, the sights may move, causing them to be misaligned. Therefore, the Marine is constantly re-aligning the sights as pressure is applied to the trigger. Sight alignment and trigger control must be performed simultaneously to fire an accurate shot.

Breath Control

During combat, the Marine's breathing and heart rate often increase due to physical exertion or the stress of battle. The key to breath control in field firing is to stop breathing just long enough to fire an accurate shot or a series of shots.

CHAPTER 4
PISTOL FIRING POSITIONS AND GRIP

The M9 service pistol is fired from the standing, kneeling, and prone positions. Each firing position may be adapted to either a Weaver or Isosceles variation, each possessing a distinct advantage in combat. The Weaver variation stabilizes the pistol sights. The Isosceles variation manages recoil. The advantages apply whether the Marine is firing in the standing, kneeling, or prone position. The Marine must select and assume a stable firing position that provides a solid foundation for accurate shooting while meeting the demands of the combat situation

Note: The procedures in this chapter are written for right-handed Marines; left-handed Marines should reverse directions as needed.

The Pistol Firing Position

During combat, the Marine selects a firing position based on mobility, observation of the enemy, and stability.

Mobility

A firing position must provide mobility should the Marine need to move. The standing position permits maximum mobility because it is quickly assumed and easily maneuvered from and it permits lateral mobility to engage widely dispersed targets. The prone position provides limited mobility because it is the most time-consuming position to get into and out of and it lacks the lateral mobility required to engage dispersed targets.

Observation of the Enemy

A firing position should allow observation of the enemy while minimizing the Marine's exposure. During combat there can be many obstructions to a clear field of view. Terrain features such as vegetation, earth contours, and manmade structures often dictate the firing position. The prone position normally allows the least exposure, but it usually provides a limited field of view. Kneeling may provide a wider field of view, but generally provides less concealment.

Stability

A solid firing position establishes a stable foundation for target engagement and provides accurate and consistent shooting. The definition of a stable position is one in which the body is positioned so as to resist forces that cause motion (i.e., recoil and movement of the pistol sights). The standing position is the least stable firing position, while the prone is the most stable firing position. A consistent, stable position is assumed for two distinct purposes:

- Minimize the pistol's movement in order to control the sights and to deliver accurate fire on a target.
- Minimize the affects of recoil in order to recover the sights to the same area on the target.

During combat, it may be necessary to engage the same target more than once to eliminate it. If the firing position is stable, the pistol's sights should recover to the same area on the target, allowing rapid re-enactment. Distributing the body's weight to balance the position also stabilizes the pistol and allows better management of recoil. A pistol firing position is stabilized through controlled muscular tension. Because the pistol is fired without benefit of bone support, muscular tension is needed in the body to stabilize the position and the pistol sights. Muscular tension must follow these guidelines—

- A consistent amount of muscular tension is needed to hold the pistol steady so the sights are aligned with the aiming eye and the target.

- Controlled and consistent tension in the body provides resistance that the Marine uses to manage recoil and bring the sights back on target quicker. However, too much tension can cause strain or trembling.
- Muscular tension is correct when the Marine can control the pistol before, during, and after firing the shot.

Pistol Firing Grip

A proper grip is one that provides maximum control of the pistol before, during, and after firing. It stabilizes the pistol sights before firing, allows trigger control to be applied during firing, and manages recoil after firing.

Before Firing: Stabilize the Sights

To fire an accurate shot, the sights must be stabilized prior to and as the bullet exits the muzzle of the pistol. A proper grip controls the alignment of the pistol's sights and stabilizes the sights so an accurate shot may be fired. The following guidelines apply to the establishment of a proper grip—

- There must be muscular tension in the wrist and forearms. Consistent muscular tension in the wrist, forearms, and grip helps maintain sight alignment by reducing the movement in the grip that can cause movement in the sights. The grip is correct when it allows the sights to be naturally aligned to the aiming eye.
- When establishing a two-handed grip, equal pressure must be applied with both hands. Consistent, equal pressure from both hands stabilizes the sights and allows them to be aligned and level with respect to the aiming eye.
- The hand grips as high on the back strap as possible, bringing the centerline of the bore as close as possible to a parallel line with the bones in the forearm.

During Firing: Allow Trigger Control

The Marine's grip provides a foundation for the movement of the trigger finger. The trigger finger applies positive pressure on the trigger as an independent action, completely free of the other muscles of the gripping hand. The Marine should not apply excessive pressure to the web of the hand touching the pistol's back strap because excessive pressure interferes with the manipulation of the trigger by the trigger finger.

After Firing: Manage Recoil

Once a shot is fired, the pistol recoils, disturbing alignment of the sights. A proper grip facilitates a quick recovery from recoil so the sights quickly return to the same area on the target. The Marine's grip determines the following during recoil:

- The amount the muzzle climbs during recoil depends on the amount of controlled muscular tension in the grip and wrists applied to stabilize the pistol and create consistency in resistance to recoil. Controlled muscular tension allows the sights to recover consistently back on target within a minimum amount of time.
- Firm, equal pressure must be applied to the grip with both hands to ensure that the pistol does not slip during recoil. An improper grip or lack of controlled muscular tension causes the pistol to move after the shot is fired, disrupting sight alignment and requiring the Marine to re-establish his grip.
- Locking out the wrist of the firing hand, similar to executing a punch, helps provide resistance to recoil and speed recovery. The elbows should be at a position slightly less than locked out to help absorb the recoil and aid in recoil management.

Withdrawing the Pistol From the Holster

The firing grip is not established in the holster, however, a proper firing grip can be assumed quickly if the pistol is withdrawn from the holster correctly. The pistol should be withdrawn from the holster in one continuous, fluid motion:

- Place the heel of the left hand at the center of the torso with the fingers extended toward the target. At the same time, unfasten and release the D-ring with the right hand. See figure 4-1.

Figure 4-1. Withdrawing the Pistol From the Holster—Step 1.

Note: The left hand is placed on the torso in a position that allows a two-handed firing grip to be established in a minimum amount of movement.

- Place the right thumb on the forward edge of the lower portion of the holster and the fingers around the back edge of the holster, keeping the trigger finger straight. See figure 4-2.

Note: The left hand may assist in holding the holster flap up.

Figure 4-2. Withdrawing the Pistol From the Holster—Step 2.

- Slide the hand up the holster until the fingers come in contact with the pistol grip. At the same time, keep the thumb above the pistol to guide the holster flap up. See figure 4-3 on page 4-4.
- Grasp the pistol grip with the fingers and draw the pistol straight up. Continue withdrawing the pistol while moving the thumb to a position on the safety. See figure 4-4 on page 4-4.

Note: This hand placement allows a firing grip to be established once the thumb disengages the safety. Any adjustments made to the firing grip after the safety is disengaged should be minor.

Figure 4-3. Withdrawing the Pistol From the Holster—Step 3.

Figure 4-4. Withdrawing the Pistol From the Holster—Step 4.

- Once the muzzle clears the holster, rotate the muzzle forward to clear the body. Ensure the muzzle is pointed in a safe direction.
- Establish a two-handed grip on the pistol by joining the left hand with the right hand in the front of the torso. See figure 4-5.

Caution

Ensure the muzzle does not cover the left hand when establishing the two-handed grip.

Weaver and Isosceles Variations

The size of the target, distance to the target, time, and type of engagement needed (i.e., two shots, single precision shot) determine whether to fire in the Weaver or Isosceles variation. The Weaver and Isosceles variations each consist of three firing positions: standing, kneeling, and prone. The firing position chosen is based on the combat situation and the Marine's body configuration, and

Figure 4-5. Withdrawing the Pistol From the Holster—Step 5.

it must permit balance, control, and stability during firing. In addition, there is a specific firing grip that supports each position and the combination of the grip and the body configuration is key to establishing the variation. See table 4-1 for a list of advantages and disadvantages of both the Weaver and the Isosceles variations.

At longer ranges, the target is smaller, a more precise shot is required to eliminate the target, and any small movement moves the sights off the target; therefore, the pistol must be steadied. Although the Weaver variation is effective at any distance, some aspects of the position make it more effective for long range or precision shots on small or partially exposed targets; i.e., sight alignment and sight picture are easier for stability of hold.

The Isosceles variation is effective at any distance, however, some aspects of the position make it more effective for close range engagements. When confronted with a target, the natural physical reaction is to face the target and push out with the arms. This makes the Isosceles variation advantageous for quick engagements at close range. When a target is at close range, it must be engaged quickly before it engages the Marine. The management of recoil is a bigger factor in close-range engagements because it is more likely that multiple shots will be fired to eliminate the target and the sights have to recover quickly back on target. However, at close ranges, the target is larger so stability of hold is not as important because it is easier to hold the sights on target and sight picture is not as critical.

Standing Position

The standing position is the most often employed position during a pistol engagement due to the short distance of the engagement and the nature of combat. When properly assumed, the standing position provides a stable base for firing, a clear field of view, and excellent mobility. The standing position can be adapted to either the Weaver variation or the Isosceles variation.

Weaver Standing Position

The key to successful employment of the Weaver variation is the body's angle to the target and the push-pull pressure applied to the grip. See figure 4-6.

Table 4-1. Advantages and Disadvantages.

Variation	Advantages	Disadvantages
Weaver	Additional balance, control, and stability of hold during firing due to placement of the arms (left arm bent, pistol is in close to the body). Easier to maintain sight picture because the pistol's foundation is steady.	Recoil has a greater impact due to the hand placement on the pistol (some of the pistol grip is exposed and pressure is applied in two different directions around the pistol). Recovery of the sights back on target may take longer since recoil is affected.
Isosceles	Recoil has a lesser impact because muscular tension and grip pressure are evenly distributed around the pistol. Allows quicker recovery of the sights on target.	Stability of hold is degraded due to the pistol being further from the body without support. Since it is harder to steady the pistol, acquiring sight picture becomes more difficult.

To assume the Weaver standing position—

- Face the target and make a half turn to the right, keeping the pistol oriented toward the target. This orients the body at approximately a 40 to 60 degree angle oblique to the target. The shoulders are angled to the target, the left shoulder forward of the right. The feet are about shoulder-width apart, the left foot forward of the right.
- Grip the pistol grip firmly with the right hand. Place the right thumb on the safety.
- Keep the shoulders at a 40 to 60 degree angle oblique to the target and raise the right arm and extend it across the body toward the target. Ensure that the right shoulder does not roll forward or turn toward the target.
- Extend the left arm to the target, bending the left elbow to join the left and right hands.

Figure 4-6. Weaver Standing Position.

The left elbow is inverted and tucked in toward the body so the left arm supports the pistol.

Note: The angle of the body determines how much the elbow bends.

- Establish a two-handed firing grip in the Weaver variation. See figure 4-7.
- Place the palm of the left hand over the front of the right hand so the palm covers the curled fingers of the right hand. The trigger guard should rest in the "V" formed by the left thumb and forefinger. The knuckles of the left hand are just outboard of the trigger guard. A portion of the pistol grip is exposed.
- Rest the trigger finger naturally, straight and outside of the trigger guard, so the finger can be moved quickly and easily to the trigger.
- Rest the left thumb against the receiver so that both thumbs are on the left side of the pistol. Once the safety is disengaged with the right thumb, the left thumb is placed over the right thumb and positive pressure is applied to hold the right thumb in place.
- Apply rearward pressure with the left hand and forward pressure with the right hand to achieve a "push-pull" grip. Isometric tension (push-pull) stabilizes the pistol during firing.

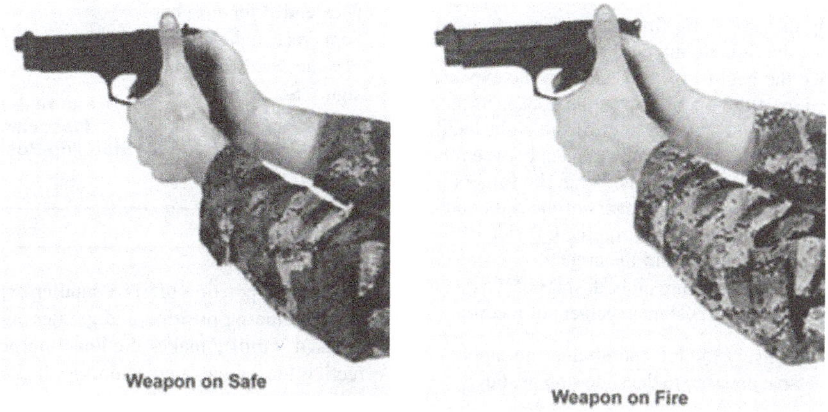

Figure 4-7. Weaver Grip.

- Lean forward slightly and apply muscular tension throughout the body to stabilize the position and manage recoil. The muscular tension in the upper body is not symmetrical due to the "push-pull" tension applied on the grip.
- Keep the head erect so the aiming eye can look through the sights.

Isosceles Standing Position

The key to the Isosceles variation is that the body is squared to the target and equal pressure is applied on the pistol from the grip. To assume the Isosceles standing position—

- Face the target with feet approximately shoulder width apart. The shoulders are squared to the target.
- Establish a two-handed firing grip in the Isosceles variation. See figure 4-8.

Figure 4-8. Isosceles Grip.

- Grip the pistol grip firmly with the right hand. Place the right thumb on the safety.
- Place the heel of the left hand on the exposed portion of the pistol grip in the pocket formed by the fingertips and heel of the right hand. There should be maximum contact between the pistol grip and the hands. Wrap the fingers of the left hand over the fingers of the right hand. Ensure both thumbs rest on the left side of the pistol and point toward the target.
- Apply equal pressure on both sides of the pistol to allow for the best management of recoil.

  Note: Ensure the left thumb does not apply excessive pressure to the slide stop or slide.

- Rest the trigger finger naturally, straight and outside of the trigger guard, so the finger can be moved quickly and easily to the trigger.

  Notes: Index finger of the left hand may or may not rest on the front of the trigger guard.

  Marines with large hands must ensure that their right thumb does not rest on the slide stop, preventing reliable pistol operation.

- Elevate and extend the arms toward the target.
- Roll the shoulders forward and shift the body weight slightly forward to stabilize the position and better manage recoil. The left foot may be slightly forward of the right foot to balance the position. There should be an equal amount of muscular tension on both sides of the body to best manage recoil.
- Tuck the head between the shoulders; the head is extended forward but kept erect so the aiming eye can see through the sights. See figure 4-9.

Figure 4-9. Isosceles Standing Position.

Kneeling Position

The kneeling position offers a smaller exposure than the standing position and greater stability. Increased stability makes the kneeling position effective for longer range shooting. It does not, however, offer as much mobility for quick reaction as the standing position. The kneeling position can be quickly assumed and allows firing from various types of cover. Depending on the cover and the need for observation, the kneeling position may be adapted to a high kneeling, a medium kneeling, a low kneeling, or a two-knee kneeling position. The kneeling positions can be adapted to either the Weaver variation or the Isosceles variation.

Weaver Kneeling Position

The advantage of the Weaver variation of the kneeling position is that it provides bone support due to the left elbow's placement on the knee. The Weaver variation further enables firing from the side of cover while exposing less of the body to a threat. To assume the Weaver kneeling position, the following steps are basic to all adaptations:

- Make a half turn to the right, drop the right foot back or step forward with the left foot, and place the right knee on the deck. The body is positioned at a 40 to 60 degree angle oblique to the target.
- Blade the shoulders at a 40 to 60 degree oblique angle to the target, the left shoulder forward of the right.
- Extend the arms toward the target.
- Bend forward at the waist to better manage recoil.
- Place the flat part of the upper left arm, just above the elbow, in firm contact with the flat surface formed on top of the bent knee. The point of the left elbow extends just slightly past the left knee. However, depending on the need for stability or observation of the enemy, the elbow does not have to rest on the knee.

The following steps provide specifics for each adaptation:

- High kneeling: the toes of the right foot are curled and in contact with the deck or the inside of the foot may be in contact with the deck. Depending on the need for observation, the buttocks may or may not rest on the right heel. The left leg is bent at the knee; the shin straight up and down. The left foot is flat on the deck. See figure 4-10.

Figure 4-10. Weaver High Kneeling.

- Medium kneeling: the right ankle is straight with the foot stretched out and the bootlaces in contact with the deck. The left leg is bent at the knee, the left foot flat on the deck. The right shin may be angled to the body to create a tripod of support for the position. See figure 4-11.

Figure 4-11. Weaver Medium Kneeling.

- Low kneeling: the right ankle is turned so the outside of the foot is in contact with the deck and the buttocks are in contact with the inside of the foot. The right shin may be angled to the body to create a tripod of support for the position. See figure 4-12.

Figure 4-12. Weaver Low Kneeling.

- Two-knee: drop both knees onto the deck. The toes may be curled to get into and out of the position quickly. Depending on the need for observation of the enemy, the buttocks may or may not rest on the heels. See figure 4-13.

Isosceles Kneeling Position

The advantage of the Isosceles variation is that it enables the Marine to fire over the top of cover while exposing less of the body to a threat. To assume the Isosceles kneeling position, the following steps are basic to all adaptations:

- Drop the right foot back or step forward with the left foot and place the right knee on the deck.
- Square the shoulders to the target.
- Extend the arms toward the target.
- Lean forward with the shoulders rolled forward and the head tucked between the shoulders to better manage recoil.

Figure 4-13. Weaver Two-Knee Kneeling.

The following steps provide specifics for each adaptation:

- High kneeling: the toes of the right foot are curled and in contact with the deck or the inside of the foot may be in contact with the deck. Depending on the need for observation, the buttocks may or may not rest on the right heel. The left leg is bent at the knee, the shin is straight up and down. The left foot is flat on the deck. See figure 4-14.

Figure 4-14. Isosceles High Kneeling.

- Medium kneeling: the right ankle is straight with the foot stretched out and the bootlaces in contact with the deck. The left leg is bent at the knee, the left foot is flat on the deck. The right shin may be angled to the body to create a tripod of support for the position. See figure 4-15.

Figure 4-15. Isosceles Medium Kneeling.

- Low kneeling: the right ankle is turned so the outside of the foot is in contact with the deck and the buttocks are in contact with the inside of the foot. The right shin may be angled to the body to create a tripod of support for the position. See figure 4-16.

Figure 4-16. Isosceles Low Kneeling.

- Two-knee: drop both knees onto the deck. The toes may be curled to get into and out of the position quickly. Depending on the need for observation of the enemy, the buttocks may or may not rest on the heels. See figure 4-17.

Figure 4-17. Isosceles Two-Knee.

Prone Position

The prone position is easily assumed, stable, and presents a small target to the enemy. Since the prone position places most of the body on the deck, it offers great stability for long range shooting. However, it is the least mobile of the firing positions and may restrict the field of view. The prone position can be adapted to either the Weaver variation or the Isosceles variation.

Weaver Prone Position

The Weaver variation of the prone position produces a cocked leg position by angling the body to the target and cocking the leg to support the position. The Weaver prone position is ideal for firing from behind cover. See figure 4-18.

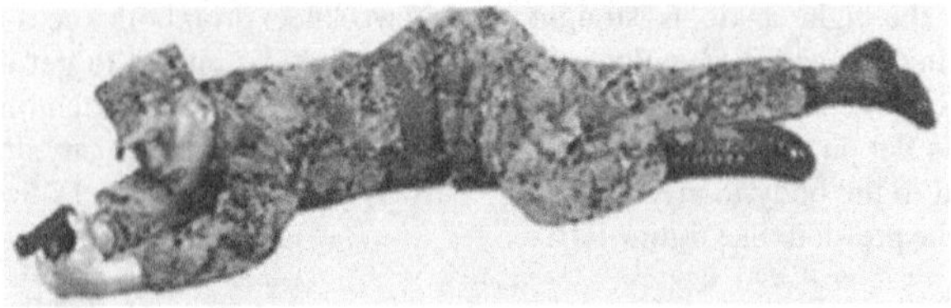

Figure 4-18. Weaver Prone Position.

To assume the Weaver prone position, perform the following steps:

- Face the target and make a half turn to the right (this places the body at a 40 to 60 degree oblique to the target). Grip the pistol in the right hand, placing the pistol in a position that facilitates control of the weapon. Ensure the pistol is pointed in a safe direction and does not cover any portion of the body.
- Move the body to the deck by using either the squat or drop method, keeping the body at a 40 to 60 degree oblique to the target.

  Squat Method
  - Squat down and place the left hand on the deck.
  - Kick both feet backward and come down on the right side of the body with the right arm extended toward the target.
  - Ensure the pistol does not cover the body or the left hand. See figure 4-19.

  Drop Method
  - Drop to a kneeling position.
  - Place the left hand on the deck in front of the body, push the pistol out toward the target.
  - Roll the right side of the body onto the deck.
  - Ensure the pistol does not cover the body or the left hand. See figure 4-20 on page 4-12.
- Bring the left knee up to support the firing position and to raise the diaphragm off the deck so as not to interfere with breathing. The inside of the knee rests on the deck. The knee is drawn up to provide maximum stability for the position.
- Establish a two handed-firing grip on the pistol.

Figure 4-19. Squat Method.

- Place the left elbow on the ground for stability. For maximum stability, strive to keep the grip firmly placed on the deck.
- The head may rest against the right arm so the pistol sights can be aligned. The head may be canted as long as the aiming eye can look directly through the sights. Strive to keep the pistol sights as level as possible while acquiring sight alignment.

To make minor increases in elevation, keep the left hand in place and firmly on the deck and raise the right hand to achieve the desired elevation. However, contact between the right and left hands must be maintained to stabilize the pistol. See figure 4-21 on page 4-12. (There is a tradeoff between obtaining the needed elevation and losing stability, so the Marine must strike a balance between the two.)

Figure 4-20. Drop Method.

Figure 4-21. Increasing Elevation (Minor Adjustments).

Isosceles Prone Position

The Isosceles variation of the prone position produces a straight leg position. See figure 4-22.

To assume the Isosceles prone position:

- Stand facing the target. Grip the pistol in the right hand, placing the pistol in a position that facilitates control of the weapon. Ensure the pistol is pointed in a safe direction and does not cover any portion of the body.
- Move the body to the deck by using either the squat or drop method.

  Squat Method
  - Squat down and place the left hand on the deck.
  - Kick both feet backward and come down on the right side of the body with the right arm extended toward the target.
  - Ensure the pistol does not cover the body or the left hand. See figure 4-23.

  Drop Method
  - Drop to a kneeling position.
  - Place the left hand on the deck in front of the body.

Figure 4-22. Isosceles Prone Position.

Figure 4-23. Squat Method.

- ❍ Push the pistol out toward the target, and roll the right side of the body onto the deck.
- ❍ Ensure the pistol does not cover the body or the left hand. See figures 4-24.
- Establish a two-handed firing grip on the pistol.
- Spread the legs to a position that provides maximum stability. The insteps of both feet may be flat on the deck or the toes may be curled and dug into the deck.
- Keep the pistol sights as level as possible while acquiring sight alignment. Keep the head in a position to allow the aiming eye to look directly through the sights.
- When wearing a helmet, the head may be canted slightly and rest against the right arm to push the helmet from the eyes so the sights can be aligned. Likewise, the pistol may be canted outboard to allow the aiming eye to look directly through the sights.

To make minor increases in elevation, keep the left hand in place and firmly on the deck and raise the right hand to achieve the desired elevation. However, contact between the right and left hands must be maintained to stabilize the pistol. See figure 4-25 on page 4-14. (There is a tradeoff between obtaining the needed elevation and losing stability, so the Marine must strike a balance between the two.)

Natural Body Alignment

The body must be properly aligned to the target so the sights fall naturally on the target when the pistol is presented. It takes a combination of body alignment and consistent muscular tension to

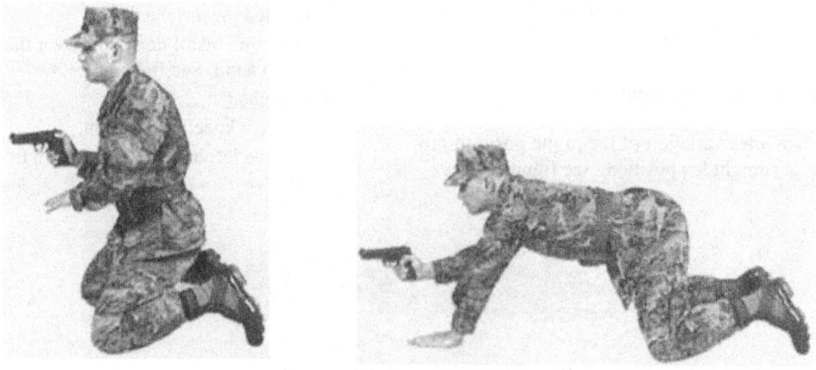

Figure 4-24. Drop Method.

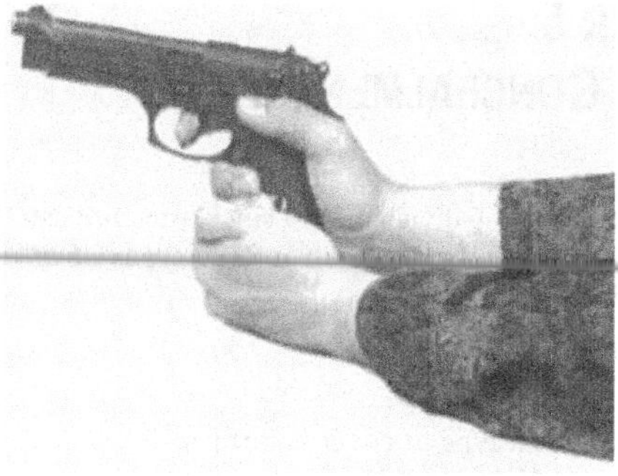

Figure 4-25. Increasing Elevation (Minor Adjustments).

ensure the sights fall naturally to the same area of the target every time the pistol is presented. The Marine can execute the following to check natural body alignment and to ensure the sights center on the aiming area:

- Orient the body to a target and establish a variation of the standing position and a two-handed firing grip on the pistol. Aim in on the target.
- Close the eyes and take a deep breath.
- Open the eyes and see where the pistol sights are in relation to the target. If the pistol sights are right or left of the target, move the feet to adjust the position right or left.

Note: Do not force the sights onto the target area by moving the arms; this increases the muscular tension on one side of the body, disturbs balance, and makes recoil harder to manage.

If the sights are significantly out of alignment when the pistol is at eye level, it may be an indication of a poor grip. When the grip is correct (to include the muscular tension in the grip, wrist, and forearms), the sights should align to the point that only minor adjustments are needed to align the sights to the aiming eye.

Repeat the preceding steps as necessary. Body alignment and muscular tension are correct when the sights are naturally placed in the same area on the target every time the Marine aims on the target.

Chapter 5
Use of Cover and Concealment

On the battlefield, a firing position that allows maximum observation of the enemy as well as cover and concealment is a necessity. A good position provides a solid foundation for the pistol, maximizes the use of cover to provide protection from enemy fire, allows mobility, and provides observation of the enemy. Where possible, the cover should be used to provide additional support for the position.

When contact with the enemy is made, it is important to seek cover as quickly as possible. Cover is anything that provides protection from enemy fire. Cover should be, at a minimum, thick enough to stop small arms fire and high enough to protect the Marine when firing from behind it. The effective use of cover enables engagement of enemy targets while affording protection from enemy fire. Cover can also be effectively used to conceal the Marine from enemy view while searching for targets.

Concealment is anything that hides a Marine from enemy view; however, it might not afford protection. Concealment can be provided by brush, trees, etc.

> *Note:* The procedures in this chapter are written for right-handed Marines; left-handed Marines must reverse directions as needed.

Cover Materials

Natural cover (rocks, logs, rubble, etc.) is best because it is hard to detect. But any material (including buildings, structures, etc.) that may protect an individual from small arms fire can be used for cover. Some common types of cover material are as follows.

Dirt

The best type of cover is dirt packed to a minimum thickness of 18 inches.

Cinder Blocks

Cinder blocks used as foundations for houses or walls can be used for cover, but they can be penetrated. They are also brittle and can shatter upon impact from small arms fire, which can cause injury by secondary fragmentation.

Trees, Logs, and Telephone Poles

Wood is a relatively dense material; therefore, it offers good covering protection since bullets have a tendency to fragment as they penetrate. Live trees have a greater resistance to bullet penetration than dead wood. Wood that has been treated with creosote, such as telephone poles and railroad ties, offers better protection from projectiles than untreated wood, but it still does not ensure protection from small arms fire.

Sandbags

Sandbags can be used for cover. However, they should always be packed tightly and tamped down to increase their density. If loosely filled or moist, a bullet can more easily penetrate the sandbag. Doubling or overlapping sandbags also increases their protective qualities.

Considerations for Firing From Cover

Adjusting the Firing Position

Cover can provide additional support for the firing position. The firing position is adjusted to ensure stability, mobility, and observation of the enemy.

The Marine chooses a firing position based on his height in relation to the height of the cover. The firing position must minimize exposure to the enemy but allow observation of the area.

Although the cover provides additional support, the Marine continues to apply the same amount of muscular tension in the grip, wrist, and forearms. Muscular tension is still necessary to stabilize the pistol sights and manage recovery.

Because the sights are higher than the muzzle of the pistol, the Marine must ensure that the muzzle of the pistol clears the cover as he obtains sight alignment/sight picture on the target. The closer the Marine is to the cover, the easier it is to ensure the muzzle clears the cover.

Applying the Weaver and Isosceles Variations

The type of cover may dictate which variation of a firing position will be the most effective. The position should provide the Marine with the maximum amount of stability and control and allow the Marine to manage recoil effectively in order to recover on target.

Weaver Variation

The Weaver variation exposes less of the body from behind cover due to the angle of the body. The Weaver variation may be better suited for firing from behind the right or left side of cover. For example, the Weaver prone is ideal for firing from behind a log.

Isosceles Variation

The Isosceles variation is good for firing over the top of cover (e.g., a window). The Isosceles prone is ideal for firing from behind narrow cover (e.g., a telephone pole).

Keeping the Body Behind Cover

Avoid inadvertent exposure of any part of the body. Be especially aware of the top of the head, elbows, knees, or any other body part that may extend beyond the cover.

Log or Curb

When firing from behind a log or curb, the Marine must present the lowest possible silhouette and may use the log or curb for maximum support of the position. For maximum protection, muzzle clearance is kept as close as possible to the top of the log or curb. The Marine may fire from either the side or the top of the log, depending on cover and concealment (see fig. 5-1). The Marine fires over the top of a curb when it is used for cover.

Figure 5-1. Firing Around a Log.

Wall or Barricade

Firing is done from either the side or over the top of a wall or barricade. See figure 5-2.

Figure 5-2. Firing From a Barricade.

Window

If the Marine has not been detected by the enemy, he should use the side of the window or the window sill for support. Ideally, it is best to fire from the corner of the window sill when using the window for support. If there is little chance for detection or the shot can be made without support, the Marine should remain back and to the side of the window opening so the pistol does not protrude and his body is concealed by the shadows/darkness of the room. If the Marine is positioned too close to the window, his body provides a silhouette to the enemy.

Vehicle

In many combat situations, particularly in urban environments, a vehicle may be the best form of cover. When using a vehicle for cover, the engine block provides the most protection from small arms fire. The Marine establishes a position behind the front wheel or front door jamb so the engine block is between him and the target. See figure 5-3.

From this position, the Marine fires over the hood of the car or underneath the car from behind the wheel. At the back of the car the axle and the

Figure 5-3. Firing Over the Hood of a Vehicle.

wheel provide the only cover. If the Marine must shoot from the back of the car, he must position himself directly behind the wheel as much as possible. See figure 5-4.

Figure 5-4. Firing From the Back of a Vehicle.

Providing Support for the Position and the Pistol

Support helps stabilize both the firing position and the pistol and enable the Marine to maintain sight alignment and sight picture.

The forearms or hands can contact the support to stabilize the pistol. The Marine may rest the pistol on or against the support as long as the support does not interfere or affect the pistol's cycle of operation. See figure 5-5.

Figure 5-5. Hand Resting on Support.

When firing over the top of cover, the Marine can establish a supported position and stabilize the position by resting the trigger guard or the magazine on the cover. The pistol may be pushed up against the support so the "V" formed by the receiver and the front of the trigger guard rests firmly against the support. See figure 5-6.

Figure 5-6. Trigger Guard Resting Against Support.

When firing around the sides of cover, the Marine can establish support and stabilize the position by placing the back of the hand or arm against the cover. Avoid placing the slide of the pistol against the cover because it can interfere with the pistol's cycle of operation. However, the pistol can be canted and placed against the cover so the trigger guard or the "V" formed between the receiver and trigger guard rests against the support. This position enables the Marine to expose less of himself to the enemy. See figure 5-7.

Figure 5-7. Hand Resting Against Side of Support.

When using a vehicle for cover, the Marine can establish additional support for the pistol by positioning himself in the car behind the door jamb (frame of door) and placing his hands or pistol against the "V" formed by the open door and door frame. See figure 5-8.

When shooting from the left side of cover, the Marine still uses his right hand and eye. See figure 5-9. He may have to cant his head and the pistol to the left to establish sight alignment. For right-handed Marines, shooting from the left side of cover may expose more of the Marine to the enemy than shooting from the right side.

Figure 5-8. Firing From Behind the Door Jamb of a Vehicle.

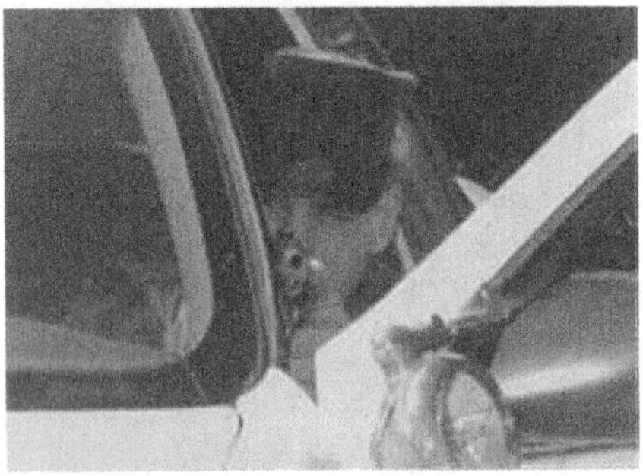

Figure 5-9. Firing From the Left Side of Cover.

Changing Positions

If the Marine has been firing from cover and has to reload or clear a stoppage behind cover he should attempt to resume firing from a different position. The enemy is aware of the Marine's current position and will be ready to engage him once he re-appears.

Moving

In combat, the Marine must be constantly aware of his surroundings and the available cover should enemy contact occur during movement from one position to another.

When moving from cover to cover, the Marine selects the next cover location and plans the route of movement before leaving his present position. This is done by quickly looking from behind cover to ensure the area is clear, ensuring the head and eyes are exposed for as short a time as possible. If necessary, the Marine should conduct a Condition 1 reload before moving from cover. Once the Marine is committed to moving, he must focus on the move until cover is reassumed.

Supported Firing Positions

During combat, the Marine may not have the time to assume a perfect firing position. He must know instinctively that his position is correct rather than follow a regimented sequence of movements to ensure its correctness. With training, the Marine can assume stable firing positions quickly and instinctively by incorporating the use of cover for support. Support provides foundations for the firing position; which, in turn, provides support for the pistol. To maximize the support provided by the position, the firing position should be adjusted to fit or conform to the shape of the cover.

A supported firing position minimizes exposure to the enemy, maximizes the stability of the pistol and protection from fire, and provides observation of the enemy. Any stable support may be used (e.g., logs, sandbags, walls). The surrounding environment dictates the support and position.

The size, distance to the target, and time affect the need for stability and recovery in the selection of a supported firing position. For example, if the target is a great distance from the Marine, he may sacrifice some of his ability to manage recoil in order to assume a supported firing position that provides him the additional stability needed to fire accurately at long range. But, recovery may be more important for the Marine if he must fire multiple shots on target quickly; therefore, he may sacrifice some stability in his supported firing position in order to engage a target with multiple shots.

Supported Prone

The supported prone position presents the lowest silhouette and provides maximum protection from enemy fire. The supported prone position can be assumed behind a tree, a wall, or almost any type and size of cover. It is flexible and allows firing of the pistol from all sides. To assume the supported prone position and maximize the use of cover, the position is kept as low as possible to ensure no part of the body is exposed to the enemy. If the cover is narrow, the Marine positions his body directly behind cover and keeps his legs together. The Marine's body is in line with the pistol and directly behind the pistol. The Isosceles prone position presents a smaller target to the enemy and more body mass to absorb recoil. The Weaver prone position is ideally suited to fire from around cover (e.g., log) because of the angle of the body, but this position presents a larger target to the enemy. See figure 5-10.

Figure 5-10. Weaver Prone.

Supported Kneeling

If the prone position cannot be used because of the height of the support, the Marine may use the supported kneeling position. The kneeling position allows firing of the pistol from all sides. This position may be altered to maximize the use of cover or support by assuming a variation of the kneeling position (high, medium, low, or two-knee). The kneeling position provides more mobility than the prone position. See figure 5-11.

Caution

The Marine must not indicate to the enemy his position with his exposed knee. If the Marine changes knees while exposed to the enemy, he has indicated the direction of his next shot to the enemy.

Figure 5-11. Supported Kneeling.

Supported Standing

When use of the support is maximized, the supported standing position can be as stable as the supported kneeling or prone position. The supported standing position provides greater mobility than the other positions and usually provides greater observation of the enemy. In the standing position, the Marine must not allow the placement of his foot to indicate to the enemy his position behind cover. See figure 5-12.

Figure 5-12. Supported Standing.

Locate and Engage Targets From Behind Cover

To locate targets when behind cover or to ensure the area is clear before moving, the Marine must expose himself to the enemy. There are two techniques used to locate and engage targets from behind cover: the pie and the rollout. These techniques minimize the Marine's exposure to enemy fire while placing him in a position to engage targets or to move to another location if necessary. Both techniques are used in the kneeling and standing positions.

Pie Technique

The Weaver position is the most effective position when executing the pie technique because the position of the Marine's body minimizes exposure to the enemy. See figure 5-13.

Figure 5-13. Pie Technique.

To perform the pie technique—

- Stay behind cover while moving back and away from the leading edge of the cover. The surroundings and situation dictate the distance the Marine moves back and away from the cover. Generally, the further back the Marine is from cover, the greater his area of observation; staying too close to cover decreases the area of observation.
- Assume a firing position and lower the pistol sights enough to have a clear field of view, aiming in on the leading edge of the cover.
- Take small side steps and slowly move out from behind cover, covering the field of view with the aiming eye and muzzle of the pistol. Wherever the eyes move, the muzzle should move (eyes, muzzle, target).
- Continue moving out from cover until a target is identified or the area is found to be clear. If a target is identified, sweep the safety off, place the finger on the trigger, and engage the target.

Note: If a target is identified before moving out from cover, the pistol should be thumb-cocked and off safe before moving out.

Rollout Technique

In this technique, the Isosceles position is the most effective position when executing the rollout technique because the position of the body allows the Marine to better maintain his balance. See figure 5-14. To perform the rollout technique—

- Stay behind cover, move back, and position the body so it is in line with the leading edge of the cover, ensuring that no part of the body extends beyond cover.
- Assume firing position and come to the Ready, ensuring the muzzle is just behind cover.
- Cant the head and pistol slightly and roll the upper body out to the side enough to obtain a clear field of view. Keep the feet in place and push up on the ball of one foot to facilitate rolling out.
- Continue to roll out from cover until a target is identified or the area is found to be clear. If a target is identified, sweep the safety off, place the finger on the trigger, and engage the target.

Figure 5-14. Rollout Technique.

Note: If a target is identified before moving out from cover, the pistol should be thumb-cocked and off safe before moving out.

Combining the Pie and Rollout Techniques

In some situations, it may be necessary to utilize both the pie and rollout technique in order to search an entire area for targets (i.e., corner of a building, doorway). Changing from one technique to another may allow the Marine to minimize his exposure to the enemy and reduce the time he is exposed to enemy fire.

CHAPTER 6
PRESENTATION OF THE M9 SERVICE PISTOL

In combat, targets present themselves with little or no warning. The Marine must have the ability to react quickly and to effectively and efficiently present the pistol, whether the pistol is in the holster or at a carry. To successfully engage a combat target with the M9 service pistol, the Marine must master pistol presentation from the carries and transport, pistol presentation while assuming a firing position, and search and assess techniques.

> *Note:* The procedures in this chapter are written for right-handed Marines; left-handed Marines should reverse directions as needed.

Sight Alignment/Sight Picture

Pistol presentation drills and dry fire help the Marine achieve a consistent grip and rapid presentation and aid in quickly acquiring sight alignment and sight picture. The Marine always executes the following steps:

- Identify the target and quickly present the pistol to the target while simultaneously sweeping the safety with the thumb of the right hand. (Disengaging the safety with the right thumb ensures the trigger is not pulled before taking the pistol off safe.)
- Shift the focus from the target to the front sight to obtain sight alignment while presenting the pistol. As the front sight breaks the plane of vision, acquire the front sight and begin to apply trigger control as sight picture is acquired.
- Apply trigger pressure until the shot is fired.

Presentation From the Carries and Holsters

Presentation From the Ready

To present the pistol from the Ready, the Marine performs the following steps in sequence. When a target appears—

- Sweep the safety with the thumb of the right hand, place the trigger finger on the trigger, and raise the arms to bring the pistol to the target.

  Note: If the Marine wishes to thumbcock the pistol for a single action shot, the pistol is thumbcocked with the left thumb after the safety is swept with the right thumb. The grip of the left hand may have to be broken to thumbcock the pistol; re-establish the grip after thumbcocking.

- Acquire sight alignment and sight picture within the aiming area and apply trigger pressure until the shot is fired.

Presentation From the Alert

To present the pistol from the Alert, the Marine performs the following steps in sequence. When a target appears—

- Sweep the safety with the thumb of the right hand, place the trigger finger on the trigger, and bring the pistol to the target:
  - If the arms are straight, raise the arms to a 45-degree angle with the deck.
  - If the arms are bent, straighten the arms out toward the target.

Note: If the Marine wishes to thumbcock the pistol for a single action shot, the pistol is thumbcocked with the left thumb after the safety is swept with the right thumb. The grip of the left hand may have to be broken to thumbcock the pistol; re-establish the grip after thumbcocking.

- Acquire sight alignment and sight picture within the aiming area and apply trigger pressure until the shot is fired.

Presentation From a Holster Transport

The pistol is presented from the holster in one continuous, fluid motion.

M12 Holster

To present the pistol from the M12 holster transport, perform the following steps in sequence once the target appears:

- Place the heel of the left hand at the center of the torso with the fingers extended toward the target. (The placement of the left hand allows a two-handed grip to be established in a minimum amount of movement.) At the same time, unfasten and release the D-ring with the right hand.
- Use the right hand to place the thumb on the forward edge of the holster and the fingers around the back edge of the holster, keeping the trigger finger straight.
- Slide the hand up the holster until the fingers come in contact with the pistol grip. At the same time, keep the thumb above the pistol to guide the holster flap up.
- Grasp the pistol grip with the fingers and draw the pistol straight up. Continue withdrawing the pistol while moving the thumb to a position on the safety.

Note: Hand placement should allow the firing grip to be established once the thumb disengages the safety. Any adjustments made to the firing grip after the safety has been disengaged should be minor.

- Once the muzzle clears the holster, rotate the muzzle forward while sweeping the safety.
- Establish a two-handed grip on the pistol by joining the left hand with the right hand in front of the torso. At the same time, start to straighten the pistol out toward the target. See figure 6-1.

Figure 6-1. Presentation From the M12 Holster—Step 1.

- Continue moving the pistol toward the target and, at the same time, place the trigger finger on the trigger, acquire sight alignment and sight picture within the aiming area, and apply trigger pressure until the shot is fired. See figure 6-2.

Figure 6-2. Presentation From the M12 Holster—Step 2.

When time permits to thumbcock the pistol for a single action shot (e.g., a long-range engagement), the Marine presents the pistol from the holster by performing the following steps once the target appears:

- Withdraw the pistol from the holster.
- Sweep the safety with the thumb of the right hand.
- Join the left and right hands and thumbcock the pistol with the left thumb.
- Establish a two-handed grip on the pistol.
- Straighten the pistol out toward the target and, at the same time, place the trigger finger on the trigger, acquire sight alignment and sight picture within the aiming area, and apply trigger pressure until the shot is fired.

M7 Shoulder Holster

The M9 service pistol fits very snugly in the M7 shoulder holster because the holster was originally designed for the M1911A1 .45-caliber pistol, which has a more slim, round design, particularly around the trigger guard. A firm grip is required when holstering and withdrawing the M9 service pistol from the M7 shoulder holster. To present the pistol from the M7 shoulder holster, perform the following steps:

- Unsnap the thumb snap closure with the left hand.
- Wrap the fingers of the right hand around the pistol grip and rest the thumb on top of the inside of the holster. See figure 6-3.

Figure 6-3. Presentation From the M7 Shoulder Holster—Step 1.

- Grasp the pistol grip firmly and draw the pistol up and away from the holster while rotating the thumb in a position to operate the safety.
- Continue withdrawing the pistol until the muzzle clears the holster and rotate the muzzle toward the target. See figure 6-4.

Figure 6-4. Presentation From the M7 Shoulder Holster—Step 2.

- Sweep the safety with the thumb of the right hand while starting to punch the pistol out toward the target.
- Establish a two-handed grip on the pistol by joining the right hand with the left hand in the center of the torso. See figure 6-5.
- Continue punching the pistol out and, at the same time, place the trigger finger on the trigger, establish sight alignment and sight picture within the aiming area, and continue trigger pressure until the shot is fired.

Assault Holster

To present the pistol from the assault holster, perform the following in one, continuous motion:

Figure 6-5. Presentation From the M7 Shoulder Holster—Step 3.

- Bring the heel of the hand down on the hammer of the pistol so that it disengages the thumb break. See figure 6-6.
- Sweep the hand forward in a small circular motion, slide the hand up the holster, and bring the fingers up under the pistol grip.

Note: Depending on the type of retention strap, this circular motion releases the retention strap or pushes the retention strap forward and out of the way.

Figure 6-6. Presentation From the Assault Holster—Step 1.

- Grasp the pistol grip and draw the pistol straight up and out of the holster while establishing a firing grip and positioning the thumb on the safety to operate it. See figure 6-7.

Figure 6-7. Presentation From the Assault Holster—Step 2.

Concealed Pistol Holster

Depending on the clothing worn by the Marine, there are two methods for accessing the concealed pistol holster so that the clothing does not become an obstruction while presenting the pistol. To present the pistol from the concealed pistol holster, the following steps are performed in one, continuous motion when the target appears:

- If the Marine is wearing a short jacket/sweater (waist-level) that is buttoned or zipped—
  - Move the left hand across the body and grasp the jacket or sweater just above the holster, at the same time place the right hand on the holster below the jacket or sweater. See figure 6-8 on page 6-6.
  - Pull up on the jacket or sweater with the left hand. At the same time, slide the right hand up the holster until the fingers come in contact with the pistol grip. Keep the thumb above the pistol to ensure the jacket or sweater clears the pistol. See figure 6-9 on page 6-6.

Note: Leaning slightly forward at the waist may assist in clearing the jacket or sweater from the pistol and in removing the pistol from the holster.

- If the Marine is wearing a long jacket (below waist-level), or a short jacket or sweater that is unfastened—
  - Place the heel of the left hand at the center of the torso to hold the left side of the jacket down with the fingers extended. At the same time, place the right hand on the right side of the jacket, with the thumb underneath the right side of the jacket below the left hand. See figure 6-10 on page 6-7.
  - Keep thumb of the right hand against the body and throw the jacket back and away from the holster. See figure 6-11 on page 6-7.

Note: Keeping a weighted object (e.g., keys, extra magazine) in the right-hand pocket of the jacket assists in throwing the jacket back and away from the holster.

  - Place the right hand on the holster and slide the hand up the holster until the fingers come in contact with the pistol grip.
- Grasp the pistol grip with the fingers and draw the pistol straight up. Continue withdrawing the pistol while moving the thumb to a position on the safety.

Note: Hand placement should allow the firing grip to be established once the thumb disengages the safety. Any adjustments made to the firing grip after the safety is disengaged should be minor.

- Rotate the muzzle forward, once it clears the holster, while sweeping the safety.
- Establish a two-handed grip on the pistol by joining the left hand with the right hand in front of the torso. At the same time, start to punch the pistol out toward the target.

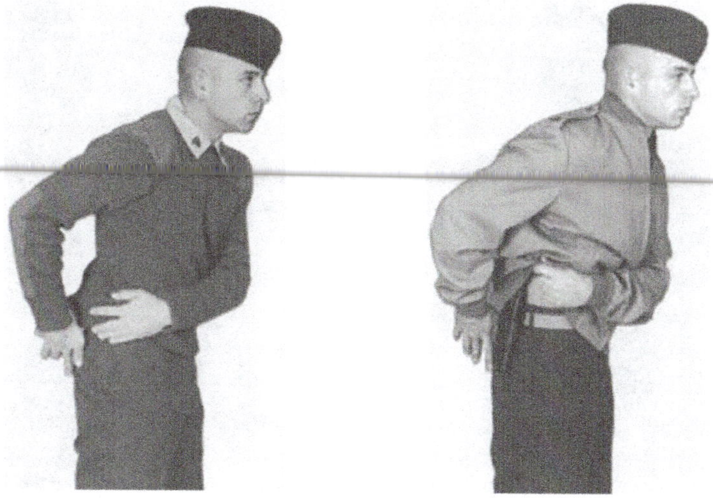

Figure 6-8. Presentation From the Concealed Pistol Holster—Step 1.

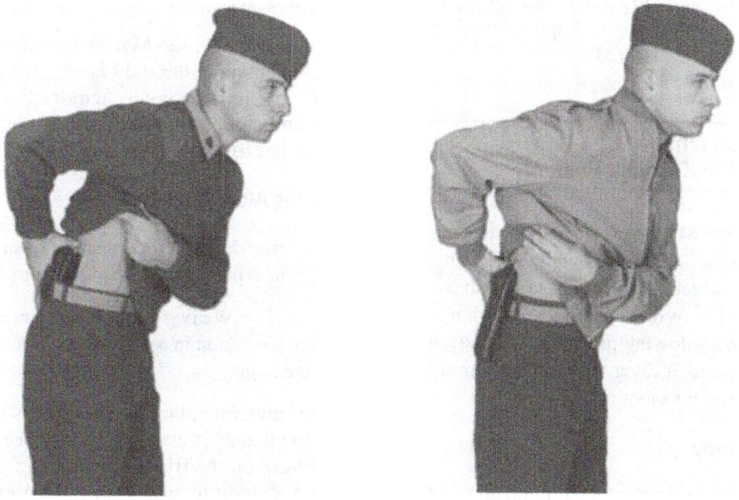

Figure 6-9. Presentation From the Concealed Pistol Holster—Step 2.

Figure 6-10. Presentation From the Concealed Pistol Holster with the Long Jacket—Step 1.

Figure 6-11. Presentation From the Concealed Pistol Holster with the Long Jacket—Step 2.

- Continue punching the pistol out and, at the same time, place the trigger finger on the trigger, acquire sight alignment and sight picture within the aiming area, and apply trigger pressure until the shot is fired.

Presentation While Assuming the Kneeling Position

In combat, the Marine must be able to quickly assume the kneeling position while presenting the pistol to the target. This usually takes advantage of cover and provides a more stable base for shooting. The following procedures are based on the Isosceles and Weaver variations of the standing to the kneeling shooting positions.

From the Ready

The Marine drops to the kneeling position while raising the arms to bring the pistol up to the target.

Note: The Weaver variation maintains the body's position in a 40 to 60 degree oblique to the target.

At the same time, the Marine sweeps the safety with the thumb of the right hand, places the trigger finger on the trigger, acquires sight alignment and sight picture within the aiming area, and applies trigger pressure until the shot is fired.

From the Alert (Arms Straight)

The Marine drops to the kneeling position while raising the arms to bring the pistol up to the target.

Note: The Weaver variation maintains the body's position in a 40 to 60 degree oblique to the target.

At the same time, the Marine sweeps the safety with the thumb of the right hand, places the trigger finger on the trigger, acquires sight alignment and sight picture within the aiming area, and applies trigger pressure until the shot is fired.

From the Alert (Close Quarters, Elbows Bent)

The Marine drops to the kneeling position while pushing the arms out toward the target.

> *Note:* The Weaver variation maintains the body's position in a 40 to 60 degree oblique to the target.

At the same time, the Marine sweeps the safety with the thumb of the right hand, places the trigger finger on the trigger, acquires sight alignment and sight picture within the aiming area, and applies trigger pressure until the shot is fired.

From the Holster Transport

The Marine withdraws the pistol from the holster while dropping to the kneeling position. The pistol should be rotated to the target by the time the knee hits the deck. The Marine sweeps the safety with the thumb of the right hand, establishes a two-handed firing grip and—

- In the Weaver variation, raises the arms to bring the pistol up toward the target.
- In the Isosceles variation, pushes the pistol out toward the target.

At the same time, the Marine places the trigger finger on the trigger, acquires sight alignment and sight picture within the aiming area, and applies trigger pressure until the shot is fired.

Presentation While Assuming the Prone Position

The prone position is generally assumed to take advantage of cover or to provide additional stability for shooting. The following procedures are executed from the standing holster transports.

Isosceles Prone

The Marine withdraws the pistol from holster. At the same time, the Marine eliminates body contact with the deck by either the squat method or the drop method. The pistol is rotated to the target as the left hand is placed on the deck. The Marine sweeps the safety off.

> *Note:* If the Marine wishes to thumbcock the pistol for a single action shot, it is done once the Marine is on the deck.

The Marine establishes a two-handed firing grip while spreading the legs a comfortable distance apart for stability. He places the trigger finger on the trigger, acquires sight alignment and sight picture within the aiming area, and applies trigger pressure until the shot is fired.

Weaver Prone

The Marine withdraws the pistol from the holster. At the same time, the Marine eliminates body contact with the deck by either the squat method or the drop method. The Marine maintains the 40 to 60 degree oblique to the target. The pistol should be rotated to the target as the left hand is placed on the deck. The Marine sweeps the safety off.

> *Note:* If the Marine wishes to thumbcock the pistol for a single action shot, it is done once the Marine is on the deck.

The Marine brings the left knee up, establishes a two-handed firing grip, and places the left elbow on the deck for stability. He places the trigger finger on the trigger, acquires sight alignment and sight picture within the aiming area, and applies trigger pressure until the shot is fired.

Search and Assess

To be successful in combat, the Marine must have the ability to assess the situation and take appropriate action following engagement. Once the Marine fires one or two rounds to engage a target, he must quickly assess the situation and the effectiveness of his engagement. This split-second assessment allows the Marine to determine the best course of action and prevents him from expending a number of rounds without assessing the results and his ability to accurately engage the target.

To search and assess, perform the following steps in sequence:

- Place the trigger finger straight along the receiver immediately after a target is engaged. Do not place the pistol on safe.
- Lower the arms just enough to look over the pistol sights and provide a clear field of view. Do not rotate the wrists in order to angle the pistol downward, this breaks the firing grip and changes the tension in the arms. Lowering the arms maintains the firing grip, keeps the sights level, and allows sight alignment and sight picture to be quickly re-established should follow-on shots need to be fired.
- Search the area by moving the head, eyes, and pistol left and right (approximately 45 degrees from center). Wherever the head moves, the muzzle moves (eyes, muzzle, target). Keeping both eyes open increases the field of view.

If the target has not been eliminated, the Marine must determine whether to re-engage the target:

- If the Marine decides to re-engage the target, the target's size, time and distance to the target, and Marine's capabilities dictate the technique used. Re-engagement techniques include re-engaging the target with a well-aimed precision shot, two shots, or offset aiming.
- If the target's distance exceeds the Marine's engagement capabilities or if the target is partially exposed and too small for the Marine to accurately engage, the Marine should not attempt to re-engage the target. The Marine may seek cover, seek out a better opportunity for engagement, or use support to better stabilize the pistol.

When it is determined that the area is clear of all enemy threat, place the pistol on safe without breaking the grip and assume a pistol carry or pistol transport.

Chapter 7
Pistol Engagement Techniques

The Marine must be able to quickly detect targets and employ the proper pistol engagement and firing techniques. The Marine must also be able to re-engage a target if the initial engagement is not successful.

> *Note:* The procedures in this chapter are written for right-handed Marines; left-handed Marines should reverse directions as needed.

Target Detection

Target indicators reveal an enemy's position. Most combat targets are detected by smoke, flash, dust, noise, or movement and are usually only momentarily visible. Target indicators are grouped into three general areas: movement, sound, and improper camouflage.

Movement

The Marine does not need to look directly at an object to notice movement. The eye is attracted to any movement, especially sudden movement. A slowly moving target is harder to detect than one with quick, jerky movements. Therefore, the ability to locate a moving target depends primarily on the speed of the object's movement.

Sound

Sound (e.g., movement, rattling equipment, talking) can be used to detect an enemy position. Sound provides only a general location, making it difficult to pinpoint a target by sound alone. However, sound can alert the Marine to the presence of a target and increases the probability of locating the target through other indicators.

Improper Camouflage

The improper use of camouflage creates three target indicators: shine, outline, and contrast with the background. The closeness of a typical pistol engagement somewhat diminishes these effects, but can still affect target detection. The Marine's ability to recognize target indicators aids in the detection of targets.

Shine

Shine is created from reflective objects such as metal, glass, pools of water, and the skin's natural oils. Shine acts as a beacon in locating a target's position.

Outline

The outline of camouflaged objects such as the body, head and shoulders, weapons, and gear can be recognizable.

Contrast With the Background

Objects contrast with a background because of differences in color, surface, and shape. The following are examples of objects contrasting with their background:

- A target wearing a dark uniform would be clearly visible in an area of snow or sand.
- Symmetrical shapes, such as helmets or rifle barrels, can be detected in a wooded area.
- Fresh soil around a fighting hole contrasts with the otherwise unbroken ground surface.

While observing an area, the Marine notes anything that looks out of place or unusual and studies it in more detail in order to increase the chances of spotting a hidden threat.

Techniques of Fire

To successfully engage a combat target with the M9 service pistol, the Marine must employ effective techniques of fire. The Marine's performance of these skills and the proper application of the fundamentals of marksmanship are critical to success in a combat situation.

Double and Single Action Firing

When the M9 service pistol is taken off safe, it is capable of firing in both a double and a single action mode.

Double Action Mode

The design of the M9 service pistol causes the first shot fired to be a double action shot.

In double action firing, two actions occur as the trigger is moved to the rear; the hammer moves to the rear, cocking the pistol, and then the hammer moves forward, firing the pistol. More pressure is required on the trigger to fire a double action shot due to the distance between the trigger and hammer and the weight of the trigger. A double action shot requires approximately 9 to 16 pounds of pressure to move the trigger rearward.

Maintaining sight alignment and sight picture are harder when firing a double action shot. Therefore, the sights will more than likely move outside the aiming area when applying trigger pressure.

Single Action Mode

In single action firing, the pistol is already cocked because the cycle of operation in the first double action shot leaves the hammer cocked to the rear. Therefore, the only action taking place as the trigger is moved to the rear is the hammer moving forward, firing the pistol. A single action shot requires approximately 4 to 6 pounds of pressure to move the trigger rearward. Therefore, the application of trigger control is easier when firing a single action shot.

To enable the first shot to be fired single action, the pistol's hammer can be manually cocked with the thumb:

Note: The pistol must be taken off safe before it can be thumbcocked.

- Use the left thumb to pull back on the hammer to cock it. This ensures that the firing grip of the right hand does not have to be broken.
- Ensure that the hammer moves all the way to the rear.
- Ensure the trigger finger remains straight along the receiver until the pistol is fully cocked.
- Re-establish the firing grip with both hands once the pistol is cocked.

Single Action or Double Action Firing Factors

The decision to fire in a single or double action mode is made rapidly. Ultimately, the decision is based on the Marine's abilities, but the decision is also based on time and accuracy. The pistol is fired in the double action mode when trigger control, sight picture, and stability of hold are not as critical for accuracy (i.e., close range, large targets). The pistol is fired in the single action mode when the fundamentals are more critical to accuracy (i.e., long range, small targets).

Time

The Marine sacrifices time to fire a single action, precision shot; but what he sacrifices in time, he gains in accuracy.

For quick engagements at close range, there may not be time to thumbcock the pistol for a single action shot. Therefore, firing the first shot in the double action mode is preferred because shots are needed on target quickly and stability of hold and sight picture are not as critical to accuracy.

When time permits and for targets at longer ranges, the pistol may be thumbcocked to place it in the single action mode to reduce the weight of the trigger and the distance the trigger must travel rearward to fire the first shot.

Distance and Size of the Target

The smaller the target, the more critical the application of the fundamentals in order to engage the target accurately. To accurately engage a small target (e.g., head shot, long-range target) it is better to thumbcock the pistol for a single action shot. Engagement of a smaller target requires additional precision because sight alignment and sight picture are more critical to accuracy.

Two-Shot Technique

During combat, the preferred pistol engagement technique is to rapidly fire more than one shot on a target to eliminate it as a threat. Two shots fired in rapid succession increase the trauma (i.e., shock, blood loss) on the target and increase the Marine's chances of quickly eliminating the threat. Therefore, two shots are most often fired in rapid succession on a target at close range.

After the pistol is fired, the muzzle climbs with the recoil of the pistol. To fire two shots, the Marine must quickly recover the sights to the same area on the target while re-acquiring sight alignment and sight picture. The proper recovery automatically brings the sights back on target following recoil. A quick recovery allows more time for the Marine to align the sights and apply trigger control to fire the next shot. Recovery begins immediately after the application of the fundamentals to bring the pistol sights into alignment with the target in preparation for firing the next shot.

The amount the muzzle climbs during recoil depends on the amount of controlled muscular tension in the grip and wrists used to stabilize the pistol and create consistency in resistance to recoil. Controlled muscular tension allows the pistol sights to recover consistently back on target within a minimum amount of time. The speed of delivery of multiple shots depends on how fast the Marine can re-acquire sight alignment. Sight recovery is determined by the following:

- The key to proper recovery is a stable firing position and proper grip.
- If the Marine's firing position is not stable, recoil forces him out of his firing position, requiring him to re-establish his position before he takes his next shot.
- An improper grip or lack of controlled muscular tension causes the pistol to move in the Marine's hand after the shot is fired, disrupting sight alignment and requiring the Marine to re-establish his grip.

Slow Fire Technique

Sight alignment becomes more critical the smaller the target and the greater the distance to the target. In these situations, the Marine does not engage the target with two rapidly fired shots because he has to slow down his application of the fundamentals in order to fire a precision shot(s). To engage small targets (i.e., partially exposed) and targets at longer ranges where precision is required, the Marine must employ the following slow fire technique:

- Thumbcock the pistol for a single action shot.
- Slow down application of the fundamentals.
- Fire one well-aimed, precision shot on target.

Re-engagement Techniques

Once the Marine has quickly assessed the situation and determined that the threat still exists, he may make the decision to re-engage the target to eliminate it as a threat. Re-engagement techniques include engaging the target with a precision shot through slow fire, two shots, or offset aiming. The time, size, and distance to the target as well as the Marine's marksmanship abilities dictate the technique used to re-engage the target.

Slow Fire Technique

If the Marine has engaged a target and the target still poses a threat, the Marine may choose to slow down his application of the fundamentals and fire a slow fire, precision shot. The placement of one well-aimed, precision shot on a designated area of the target increases the chances of eliminating the target as a threat. However, the time the Marine has to engage the target determines if he can slow down his application of the fundamentals to fire a precision shot. Firing a precision shot takes time, so the Marine will sacrifice time for accuracy.

The key to successful shot placement is the Marine's ability to slow down his performance and focus on the application of the fundamentals of marksmanship. The Marine must determine where to place shot for maximum effectiveness. In order to make this decision, the Marine must consider distance and size of the target and time. For targets at close range and within the Marine's marksmanship abilities, a precision shot is placed in the head to immediately eliminate the target as a threat. Sometimes, even at close ranges, the target may only be partially exposed; therefore, the Marine is presented with a small target and aims his sights on the portion of the target that is exposed. For targets at long ranges, the target is smaller and requires a precision shot. A precision shot may be placed in the body to add trauma to the target and increase the chances of eliminating the target as a threat.

Two-Shot Technique

If a target is accurately engaged, but does not go down, the Marine may re-engage the target with additional shots. Two additional, rapidly delivered shots increases the trauma and increases the Marine's chances of eliminating the threat.

The size and distance to the target affect how quickly two shots can be delivered on the target. The speed at which two shots are fired also depends on the Marine's marksmanship abilities and how fast he can re-acquire his front sight. However, the Marine must not compromise accuracy for speed; the key to successful target engagement is to fire only as quickly as the Marine can fire effectively.

Offset Aiming

When the Marine assesses the situation and determines his shots are not successful and not striking the target in the designated aiming area, he may employ offset aiming. Since the pistol's sights cannot be adjusted, offset aiming is applied to adjust the aiming area and to cause rounds to strike center mass. This technique should only be applied when the Marine determines that he is applying the fundamentals of marksmanship correctly. Sight picture will change as the aiming area is adjusted.

The known strike of the round offset aiming technique requires shifting the point of aim to compensate for rounds striking off target center. To effectively engage a target using this technique, the Marine must be able to see where the rounds are striking and then aim an equal distance from the center of the aiming area opposite the observed strike of the round. For example, if the rounds are striking the target high and left of center mass, aim an equal and opposite distance low and right.

Multiple Targets

If engaging more than one target at a time, the Marine adheres to the fundamentals of marksmanship and employs the techniques of multiple target engagement. The introduction of multiple targets in a combat scenario requires additional skills that must be learned and practiced if a Marine is to be successful. To be effective in combat, the Marine must be able to detect targets through identification of target indicators (identified on page 7-1), prioritize the targets, and employ multiple target engagement techniques.

Combat Mindset

Successful engagement of multiple targets requires a somewhat different mindset from single target engagement. For example, following engagement of a single target, the Marine assesses the situation. During multiple target engagements, after the first target is engaged, he must immediately engage the next target and continue until all targets have been eliminated. Because split-second decisions must be made, the development of a combat mindset is important to success on the battlefield. A combat mindset allows the Marine to control the pace of the battle rather than react to the threat.

Mental preparedness is essential to successful engagement of multiple targets, and the required mindset must be developed until it becomes second nature to the Marine. When multiple targets appear, the Marine must prioritize the targets to establish an engagement sequence. To prioritize targets and establish an engagement sequence, the Marine must be aware of the surroundings, not focus on one target, and continuously search the terrain for additional targets.

Prioritizing Targets

Once multiple targets have been identified, they must be prioritized in terms of the threat each target presents. While the fundamentals of marksmanship must still be applied, prioritizing targets and planning the engagement are just as essential to successful multiple target engagement.

Target priority is based on factors such as proximity, threat, and opportunity. It also encompasses the Marine's proficiency level since a Marine should not attempt to engage a target beyond his proficiency level or the pistol's capability (e.g., a target 100 yards away). The principal method of prioritizing targets is to determine the level of threat for each target so that all targets may be engaged in succession, starting with the most threatening and ending with the least threatening.

Prioritizing targets is an ongoing process. As the engagement proceeds, new targets may appear or previously identified targets may take cover, delaying their engagement. Therefore, the Marine must remain constantly alert to changes in target threat, proximity, and the opportunity for engagement and revise target priorities accordingly.

Engaging Multiple Targets

Once targets have been prioritized, the Marine must quickly eliminate them. It is crucial to understand and practice the physical techniques for engaging multiple targets until they become second nature. When these techniques occur as automatic responses, the Marine maintains an awareness of the complete battlefield and is able to concentrate on the mental aspects of multiple target engagement.

Engagement Methods

The fundamentals of marksmanship are critical to the development of skills to support multiple target engagement. While responsiveness is important when engaging multiple targets, the Marine's primary concern is to place accurate rounds on target. A balance must be struck between placing shots quickly on targets at close range and slowing down to place precision shots on targets at long range because all targets pose a threat and must be accurately engaged. To engage multiple targets—

- Engage every target, move from target to target.
- After all of the targets have been initially engaged, assess the effectiveness of the engagements and, if necessary, re-engage targets that still pose a threat.
- If all of the targets are eliminated, search the area for new targets.

Considerations for Field Firing Positions

The selection and use of field firing positions are critical to engagement of multiple targets. As in any firing situation, if the situation permits, the Marine should make a quick mental review of the terrain to select a firing position that provides stability, mobility, and observation of the enemy. The firing position should also provide flexibility for engaging multiple targets. The more dispersed the multiple targets, the greater the lateral movement the position must afford to engage them.

The prone position provides the maximum stability for firing well-placed shots, but engaging multiple targets from this position may require adjustment in the position from shot to shot depending on the distance between the targets and their location. Because the arms are fully extended on the ground, the prone position can be restrictive, increasing the time it will take to recover the sights onto subsequent targets. If the Marine must make an adjustment in the prone position to engage multiple targets, he must ensure that he maintains proper arm placement. Improper arm placement may affect his stability of hold and his ability to acquire sight picture. The following identifies position adjustments:

- For targets that are not widely dispersed, the Marine adjusts the position of the lower part of the body to orient the upper body toward the target without disturbing the placement of the arms. This allows the Marine to maintain stability of hold and quickly recover on subsequent targets. Moving the legs to the left orients the upper body to the right; moving the legs to the right orients the upper body to the left.
- For targets that are widely dispersed, the Marine may need to adjust his entire position to orient his body in the direction of subsequent targets. Using the left hand, push the upper body off the deck just high enough to move the body in the direction of the target. Extend the right arm toward the target and keep the muzzle pointed in the direction of the target.

The kneeling and standing positions provide an increased field of view and allow maximum lateral adjustment to engage dispersed targets. The type of adjustment the Marine must make in the kneeling or standing position in order to orient his body in the direction of each target is related to the distance between the targets. It is also related to the Marine's ability to maintain stability of hold and recover from recoil. Once the first target is engaged—

- Maintain a proper shooting position, rotate the body in the direction of the target, keeping the feet in place. In the standing position, the Marine may bend slightly at the knees while rotating the body. This allows the Marine to distribute his weight forward to provide additional stability to the position and to better control the effects of recoil.
- Adjust the entire position by either adjusting the placement of the feet or knee(s) in order to face in the direction of the new target if the targets are some distance apart.
- Ensure that as the eyes move to a new target, the head and pistol muzzle should follow.

Moving Targets

The majority of combat targets will be moving; the enemy will move quickly from cover to cover, exposing himself for the shortest time possible. Therefore, a moving target must be engaged before it disappears. The Marine must engage a moving target with the same accuracy and precision used to engage a stationary target. Practice in the proper application of leads and the fundamentals of marksmanship enable the successful engagement of moving targets.

Types of Moving Targets

Moving Target

This type of target moves in a consistent manner (i.e., walking or running) and is in continuous sight as it moves across the Marine's field of vision. However, unless the enemy is completely unaware of the Marine's presence, this type of target is not likely to present itself.

Stop and Go Target

This type of target appears and disappears during its movement due to intermittent cover. It will present itself for only a short period of time before re-establishing cover. This type of target is most vulnerable to fire at the beginning and end of its movement because it begins slow and gains momentum as it leaves cover and then slows to avoid overrunning the new cover.

Leads for Moving Targets

Once the moving target has been identified it must be engaged. To engage a moving target, the Marine must aim at a point in front of the target, which is known as a lead. Lead is the distance in advance of the target that is required to strike the target when it is moving. To achieve success, determining the amount of lead to engage a moving target must be as precise as possible. When a shot is fired at a moving target, the target continues to move during the time of the bullet's flight. For this reason, the aim must be in front of the target; otherwise, the shot will fall behind the target.

Factors Affecting Lead

Factors that affect the amount of lead are the target's range, speed, and angle of movement.

Range. There is a time lag from the time a round is fired until the round strikes at the point of aim. This time of flight could allow a target to move out of the bullet's path if the round were fired directly at the target. Time of flight increases as range to the target increases. Therefore, the lead must be increased in proportion to the distance to the target.

Speed. A greater lead is required to hit a running man than a walking man because the running man moves a greater distance during the flight of the bullet.

Angle of Movement. The angle of target movement also affects the amount of lead required for target engagement. The angle of movement across the Marine's line of sight relative to the flight of the bullet determines the type (amount) of lead.

Types of Leads

Full Lead. The target moves straight across the Marine's line of sight with only one arm and half the body visible. This target requires a full lead because it moves the greatest distance across the Marine's line of sight during the bullet's flight.

Half Lead. The target moves obliquely across the Marine's line of sight (at about a 45-degree angle). One arm and over half of the back or chest are visible. This target requires half of a full lead because it moves half as far as a target moving directly across the Marine's line of sight during the bullet's flight.

No Lead. A target moving directly toward or away from the Marine presents a full view of both arms and the entire back or chest. No lead is required. This target is engaged in the same manner as a stationary target because it is not moving across the Marine's line of sight.

Point of Aim Technique

To engage a moving target, a Marine establishes a lead using a point of aim offset aiming technique. This technique uses predetermined points of aim to sector a man-sized target vertically, halfway between center mass and the leading edge of the target (both one point of aim and two points of aim) (see fig. 7-1 on page 7-8). The same units of measure can be applied off the target for holds of additional points of aim. To use the point of aim technique to establish a lead on a man-sized moving target at various ranges, speeds, and angles of movement, the following guidelines apply. See figure 7-2 on page 7-9.

- For a target moving at a distance of 15 yards away or less, no lead is required.
- For a target walking directly across the Marines line of sight (full lead) between 16 and 25 yards away, hold 1 point of aim in the direction the target is moving.
- For a target running directly across the Marine's line of sight (full lead) between 16 and 25 yards away, hold 2 points of aim in the direction the target is moving.

Methods of Engagement

Moving targets, although difficult, can be engaged by the tracking or ambush method or a combination of the two. See figure 7-2 on page 7-9.

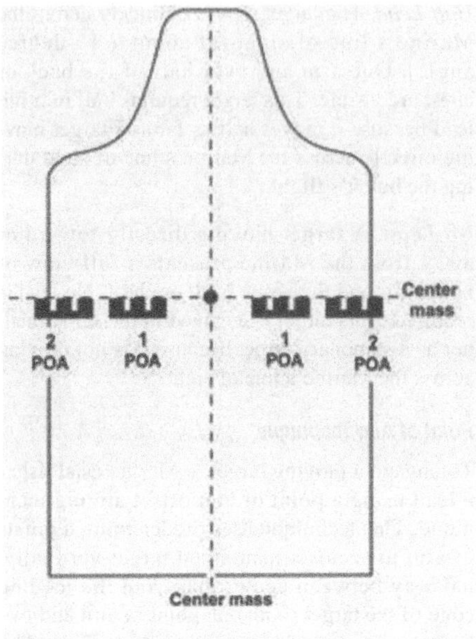

Figure 7-1. Points of Aim.

Tracking Method

In this method, the Marine "tracks" or follows the target with his front sight while maintaining sight alignment and a point of aim on or ahead of (leading) the target until the shot is fired. Sight picture is the aiming point in relation to the target while maintaining sight alignment (when a lead is established in moving target engagement, the sights are not entered on the target). To execute the tracking method—

- Thumbcock the pistol while presenting it to the target.
- Track the muzzle of the pistol through the target to the desired point of aim (lead). The point of aim may be on the target or some point in front of the target depending upon the target's range, speed, and angle of movement.
- Track and maintain focus on the front sight while applying trigger pressure and acquiring sight alignment.
- Continue tracking and applying trigger pressure while acquiring sight picture. When sight picture is established, engage the target while maintaining the proper point of aim (lead).
- Follow through so the lead is maintained as the bullet exits the muzzle. Continuing to track also enables a second shot to be fired on target.

Ambush Method

The ambush method is generally used to engage a stop and go target and when it is difficult to track the target with the pistol, such as in the prone position. With this method, the pistol is aimed at a predetermined engagement point ahead of the target and along its path, allowed to remain stationary, and fired when the target reaches the predetermined engagement point. The engagement point is based on the required point of aim (lead) to engage the target effectively. When the sights are settled, the target moves into the predetermined engagement point creating the desired sight picture. Once sight picture is established, the remaining pressure is applied on the trigger until the shot is fired. To execute the ambush method—

- Look for a pattern of exposure.
- Thumbcock the pistol while presenting it to a selected point of aim ahead of the target.
- Apply trigger pressure while obtaining sight alignment in the aiming area.
- Continue applying trigger pressure and hold sight alignment until the target moves into the predetermined engagement point and the desired sight picture is established.
- Engage the target once the sight picture is acquired.
- Follow through so the pistol sights are not disturbed as the bullet exits the muzzle.

Applying the Fundamentals of Marksmanship

The engagement of moving targets is a perishable skill that must be practiced frequently if it is to be maintained. The Marine must practice to develop the skill to calculate the point of aim (lead) and

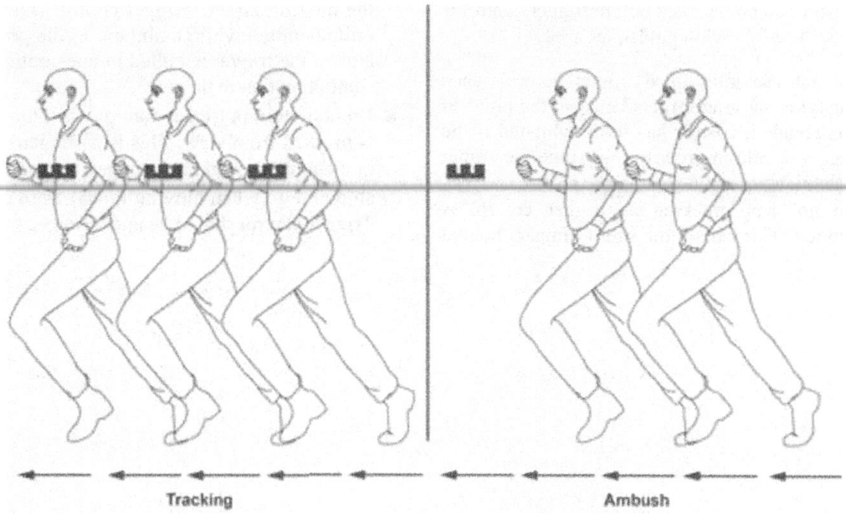

Figure 7-2. Moving Target Engagement Methods.

fire the shot while maintaining the proper point of aim (lead).

To engage moving targets using the tracking method, the Marine moves the pistol smoothly and steadily as the target moves. A stable position and firm grip are necessary to steady the pistol sights while tracking.

When using the tracking method, continue tracking the target while following through with the shot process so the point of aim (lead) is maintained as the bullet exits the muzzle. Continuing to track also enables a second shot to be fired on target if necessary. Concentration should be on continuing to track while applying the fundamentals. The fundamentals must be instinctively applied, allowing concentration on tracking the target and applying point of aim (lead).

Aiming

Sight alignment remains unchanged for accurate engagement of a moving target. The most common error when engaging moving targets is to focus on the target rather than focusing on the front sight. Sight picture is the point of aim in relation to the target while maintaining sight alignment. For both the ambush and tracking methods, sight picture is changed from the normal center mass picture, based on range, speed, and angle of movement of the target. It takes practice in moving target engagement to quickly establish the desired sight picture for a given point of aim (lead).

Breath Control

There is no difference in breath control when engaging moving targets; the breath is held to fire the shot.

Trigger Control

Trigger control is critical to firing shots while establishing and maintaining sight alignment and sight picture. Because the single action mode is the preferred method of engaging a moving target, the pistol is thumbcocked before trigger control is applied. The following guidelines apply:

- Do not use interrupted trigger control when engaging moving targets because the point of aim (lead) is lost or has to be adjusted if the trigger is held in order to re-assume the proper sight picture.
- Do not stop tracking as trigger control is applied. This causes the shot to impact behind the moving target. Trigger control should be uninterrupted while maintaining the point of aim so the trigger is pulled in one continuous, smooth motion to the rear.
- Do not interrupt trigger control when the target is masked from view. This is particularly true in the ambush method for targets that appear to stop and go when moving to and from cover. Trigger control should be uninterrupted.

CHAPTER 8
ONE-HANDED TECHNIQUES

In combat, a situation may arise where the Marine must engage a target by using only one hand. A Marine fires the pistol one-handed when—

- The Marine's other hand is occupied.
- A target presents itself so quickly and at a close range that the Marine does not have time to establish a two-handed grip and position.
- One of the Marine's hands is injured.

Note: The procedures in this chapter are written for right-handed Marines; left-handed Marines should reverse directions as needed.

Presentation and Shooting

Adjustment of the Standing Position

To fire accurately, the Marine must maintain the same stability and control of the pistol, manage recoil, and recover on target as effectively with one hand as with two. This is accomplished by adjusting the standing position as the pistol is presented to the target (see fig. 8-1).

The Marine's angle to the target determines what adjustments are made to the standing position. This adjustment is made while the pistol is being presented to the target by moving or sliding the feet to orient the body to the target. For stability and balance, the right foot may be slightly forward of the left foot. The further to the right of the Marine the target is the farther forward the right foot will be. To adjust the standing position when firing one-handed—

- Increase the muscular tension in the right arm and bend the elbow slightly to better manage the pistol.

Note: For some Marines, muscular tension in the firing arm and the position of the body in relation to the target causes them to cant the pistol inboard slightly. A slight inboard cant of the pistol and firing arm allows a natural bend in the elbow and allows a better management of recoil and recovery of the sights back on target. Any cant should be natural and not caused by excessive muscular tension. A slight cant does not affect shooting performance and should not alter the application of the fundamentals of marksmanship.

- Apply controlled muscular tension throughout the body to better manage the effects of recoil.
- Lean forward aggressively, roll the shoulders forward, and tuck the head into the shoulders.

Figure 8-1. One-Handed Presentation.

Note: The following step is done only during training to demonstrate how to stabilize the position and to simulate the Marine's other hand being occupied.

- Place the left hand in a fist on the center of the torso to increase stability. If the left arm is swinging freely it causes a corresponding movement in the pistol, reducing stability and control. Placing the left hand on the torso also ensures it is not covered by the muzzle of the pistol as the pistol is presented.

Searching and Assessing After Firing

Following one-handed target engagement, the Marine must quickly decide if it is to his advantage and if it is possible to free his occupied hand in order to place both hands on the pistol to search and assess. Whenever possible, the Marine places both hands on the pistol to increase stability and to be ready for re-engagement if necessary.

Reloading

Dry Reload

When only one hand is available, the dry reload can be performed as follows:

- Seek cover, if the situation permits.
- Press the magazine release button and allow the magazine to fall to the deck. If using the right hand, press the magazine release button with the thumb. If using the left hand, press the magazine release button with the trigger finger.
- Point the muzzle in a safe direction and position the pistol to facilitate loading the magazine. Depending on the firing position, this is accomplished using one of the following methods:

  Two-Knee Kneeling Position

  - Rotate the pistol so the magazine well faces up and the muzzle of the pistol is pointed away from the body.
  - Place the pistol between the thighs or knees and apply pressure to hold the pistol in place.
  - Withdraw a filled magazine from the ammunition pocket and insert it into the magazine well, seating it with the heel of the hand. See figure 8-2.

Figure 8-2. One-Handed Dry Reload: Two-Knee Kneeling Position.

Standing Position

- Rotate the pistol so the magazine well faces up and the muzzle of the pistol is pointed away from the body.
- Place the pistol between the thighs or knees and apply pressure to hold the pistol in place. Bend slightly at the knees to secure the pistol between the thighs.
- Withdraw a filled magazine from the ammunition pocket and insert it into the magazine well, seating it with the heel of the hand. See figure 8-3.

Kneeling Position (High, Medium, Low)

- Rotate the pistol so the magazine well faces outboard and the muzzle is pointed away from the body.
- Place the rear portion of the slide in the bend of the right knee and apply pressure with the leg to hold the pistol in place. For additional control, the Marine may drop to a two-knee kneeling position and place the pistol between his knees.
- Withdraw a filled magazine from the ammunition pocket and insert it into the magazine well, seating it with the heel of the hand. See figure 8-4 on page 8-4.

Prone Position

- The Marine places the pistol on the deck in front of him and against his body with the muzzle pointed in a safe direction. Alternately, the Marine may roll on his side and place the pistol between his knees, with the magazine well facing out.
- Withdraw a filled magazine from the ammunition pocket and insert it into the magazine well, seating it with the heel of the hand. See figure 8-5 on page 8-4.

- Grasp the pistol grip and remove the pistol from its secured location or from the deck.
- Press the slide release to allow the slide to move forward and chamber a round. If using the right hand, press the slide release with the thumb. If using the left hand, press the slide release with the index finger.

Figure 8-3. One-Handed Dry Reload: Standing Position.

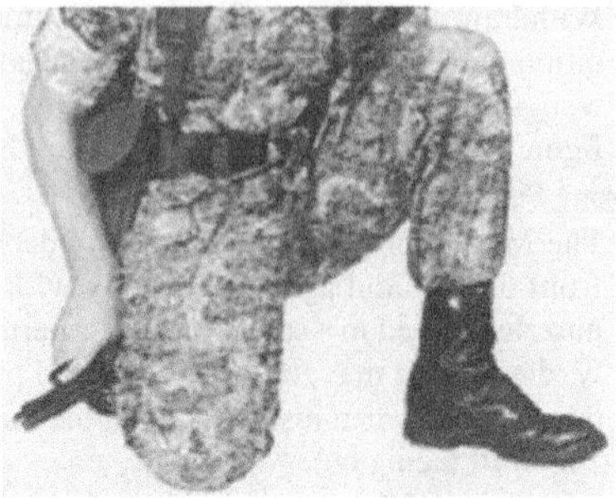
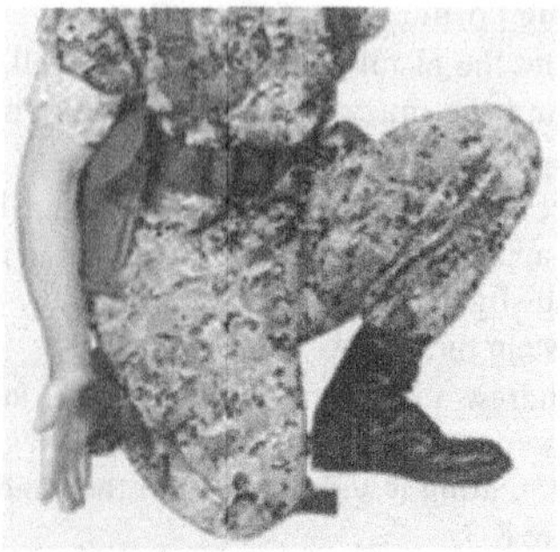

Figure 8-4. One-Handed Dry Reload: Kneeling Position.

Condition 1 Reload

When only one hand is available, a Condition 1 reload can be performed as follows:

- Point the muzzle in a safe direction and position the pistol to allow for removal of the magazine (e.g., between the thighs, in the bend of the knee, in the holster, tucked in the cartridge belt, on the deck against the body). Any position is correct if it allows the magazine well to be exposed.
- Withdraw a filled magazine from the ammunition pocket.
- Press the magazine release button and remove the magazine from the pistol. If using the right hand, press the magazine release button with the thumb. If using the left hand, press the magazine release button with the index finger.
- Insert the filled magazine into the magazine well, seating it with the heel of the hand.
- Stow the partially-filled magazine (e.g., inside the cartridge belt, in the cargo pocket).
- Grasp the pistol grip and remove the pistol from its secured location, keeping the trigger finger straight and off the trigger until ready to fire.

Remedial Action

Remedial action requires investigating the cause of the stoppage, clearing the stoppage, and returning the pistol to operation. When performing remedial action, seek cover if the tactical situation permits. Once a pistol ceases to fire, the Marine must visually or physically observe the pistol to identify the problem before it can be

Figure 8-5. One-Handed Dry Reload: Prone Position.

cleared. The steps taken to clear the pistol are based on what is observed:

- Remove the finger from the trigger and place it straight along the receiver.
- Lock the slide to the rear. To pull and lock the slide to the rear, push up on the slide stop and maintain pressure with the right thumb. (Push up on the slide stop with the left index finger if remedial action is being performed with the weak hand.)
- Secure the rear sight on the top of the cartridge belt or any other surface that provides the resistance needed to stabilize the pistol. See figure 8-6. While applying pressure on the pistol to keep the rear sight secured, push downward on the pistol in one continuous motion to lock the slide to the rear.

  Note: Ensure the pistol does not move to safe when locking the slide to the rear.

- Place the pistol in a position to observe the chamber.
- Correct the stoppage.

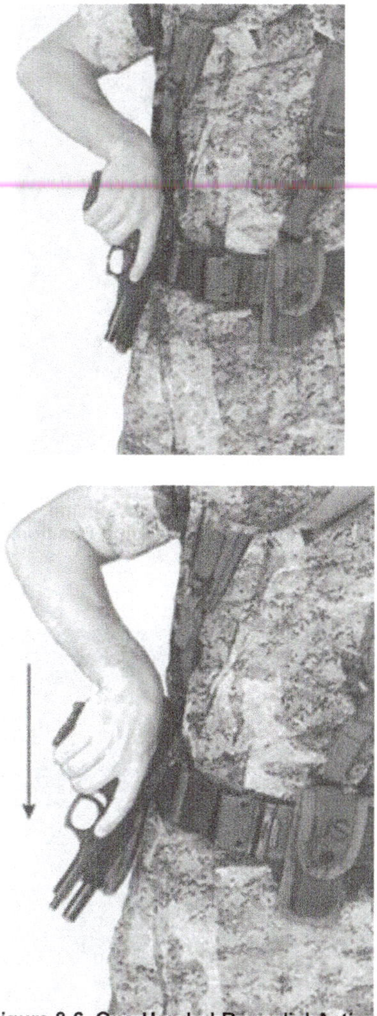
Figure 8-6. One-Handed Remedial Action.

If there is a round in the magazine but not in the chamber, release the slide and observe a round being chambered. If you do not observe a round being chambered, tap, rack, and bang to perform remedial action:

<u>Tap</u>

- Keep muzzle pointed in a safe direction and place trigger finger straight along the receiver.
- Strike the bottom of the magazine against a hard surface to ensure it is seated. In the kneeling position, strike the magazine against the thigh (see fig. 8-7 on page 8-6). In the standing position, bring the right knee up while striking magazine against the right thigh (see fig. 8-8 on page 8-6). In the prone position, strike the magazine against the deck.

Figure 8-7. Tap the Magazine Against the Knee: Kneeling Position.

Figure 8-8. Tap the Magazine Against the Knee: Standing Position.

Rack

- Rotate the pistol so the rear sight can be hooked on the top of the cartridge belt or any other surface (e.g., the edge of a table, wall, heel of the boot) that provides the resistance needed to rack the slide in a smooth, uninterrupted motion. Do not let the muzzle of the pistol cover the body. See figures 8-9 and 8-10.

Figure 8-9. Rack Slide Against Cartridge Belt.

Figure 8-10. Rack Slide Against Heel of Boot.

- Apply pressure on the pistol to keep the rear sight secured while pushing downward on the pistol to move the slide fully to the rear.
- Release pressure on the pistol to chamber the round.

Bang

- Recover the pistol on target, re-establish sight alignment/sight picture, and attempt to fire.

Presentation From the Holster With the Weak Hand

During combat, the Marine must be prepared to engage targets at any time. Therefore, the Marine must have the ability to present the pistol from the holster with his weak (left) hand if circumstances prevent him from using his strong (right) hand. He must apply the same smooth, controlled movement to present the pistol to the target. There are three methods for withdrawing the pistol from the holster with the weak hand. The Marine should practice each method to determine which works best for him.

Method One: Pistol Rotation

- Unfasten and release the D-ring with the left hand. See figure 8-11.

Figure 8-11. Method One: Release D-Ring.

- Wrap the fingers and thumb around the pistol grip so the index finger and thumb are around the base of the pistol grip. See figure 8-12.

Figure 8-12. Method One: Grasp the Pistol.

- Draw the pistol straight up until trigger guard clears the top of the holster. See figure 8-13.

Figure 8-13. Method One: Clearing Trigger Guard.

- Rotate the pistol so the magazine well faces the target and the trigger guard rests on the top of the holster. Ensure the trigger guard rests securely against the top of the holster. See figure 8-14 on page 8-8.

Figure 8-14. Method One: Rotating the Pistol.

- Maintain downward pressure on the pistol and slide the left hand around the pistol grip to establish a firing grip. See figure 8-15.

Figure 8-15. Method One: Establishing the Firing Grip.

- Grasp the pistol and continue to withdraw the pistol until the muzzle clears the holster and rotate the muzzle toward the target.
- Sweep the safety off with the thumb of the left hand while moving the pistol out toward the target. At the same time, place the trigger finger on the trigger and establish sight alignment and sight picture within the aiming area.
- Continue trigger pressure until the shot is fired.

Method Two: Hand Rotation

- Unfasten and release the D-ring with the left hand.
- Push the holster flap up and out of the way with the back of the left hand.
- Push the right hip out to facilitate grasping the pistol grip. Grasp the pistol grip between the holster flap and the pistol grip, the thumb is wrapped around the other side of the pistol grip against the backstrap. See figure 8-16.

Figure 8-16. Method Two: Grasping the Pistol.

Note: The cartridge belt may be pulled with the left hand to bring the holster closer to the body's center to facilitate grasping the pistol.

- Draw the pistol straight up while rotating the pistol so the magazine well faces inboard. Continue to remove the pistol until the trigger guard rests on the holster. See figure 8-17.
- Establish a firing grip and rotate the thumb to a position to operate the safety. See figure 8-18.
- Continue to withdraw the pistol until the muzzle clears the holster and rotate the muzzle to the target. See figure 8-19.
- Sweep the safety with the thumb of the left hand while moving the pistol out toward the

Figure 8-17. Method Two:
Trigger Guard Rests on Holster.

Figure 8-18. Method Two:
Establish Firing Grip.

target. At the same time, place the trigger finger on the trigger and establish sight alignment and sight picture within the aiming area.

- Continue trigger pressure until the shot is fired.

Method Three: Knee Placement

This method is particularly effective when firing from the two-knee kneeling position and the standing position.

Figure 8-19. Method Two:
Rotating the Muzzle.

- Unfasten and release the D-ring with the left hand.
- Wrap the fingers and thumb around the pistol grip so the index finger and thumb are around the base of the pistol grip.
- Draw the pistol straight up until the muzzle clears the holster and rotate the muzzle forward. The magazine well should face upward.
- Place the pistol between the thighs, applying enough tension to keep the pistol in place. The Marine must ensure that he does not disengage the safety while performing this motion and that the muzzle of the pistol is clear of the body. In the standing position, the Marine bends slightly at the knees to facilitate securing the pistol between the thighs.
- Remove the left hand from the pistol grip and rotate the hand around the pistol grip to facilitate establishing a firing grip.
- Grasp the pistol grip and remove the pistol from the thighs, rotating the pistol until the muzzle is pointed toward the target.
- Sweep the safety with the thumb of the left hand while starting to punch the pistol out toward the target.
- Establish a firing grip and continue to punch the pistol out toward the target. At the same time, place the trigger finger on the trigger and

establish sight alignment and sight picture within the aiming area.

- Continue trigger pressure until the shot is fired.

Transferring the Pistol From One Hand to the Other

Situations may arise in combat where the Marine must transfer the pistol to the left hand to engage a target because the right hand or arm is injured. To safely transfer the pistol from one hand to the other, perform the following steps:

Note: This procedure is written to transfer the pistol from the right hand to the left hand.

- Point the muzzle in the direction of the target and place the trigger finger straight along the receiver. See figure 8-20.

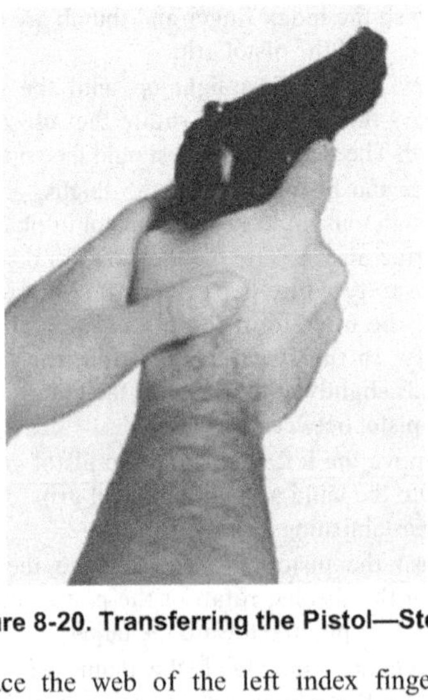

Figure 8-20. Transferring the Pistol—Step 1.

- Place the web of the left index finger and thumb under the backstrap of the pistol. See figure 8-21.
- Wrap the fingers of the left hand around the pistol grip while releasing the right hand grip.

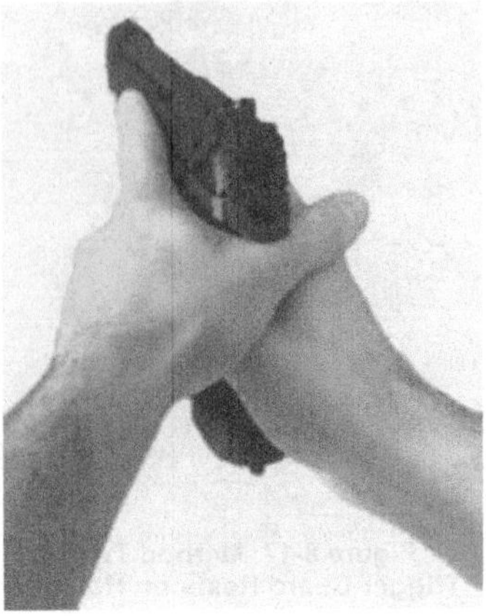

Figure 8-21. Transferring the Pistol—Step 2.

- Establish a firing grip with the left hand on the pistol. See figure 8-22.

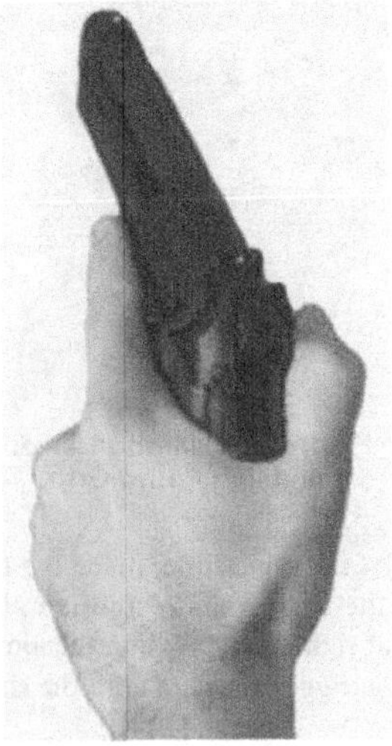

Figure 8-22. Transferring the Pistol—Step 3.

CHAPTER 9
ADVANCED TECHNIQUES

Some tactical situations are quite complex and require Marines to apply advanced pistol engagement techniques in order to effectively engage the enemy and survive. The ability of the Marine to rapidly shoot on the move and to turn and fire will be critical to a successful engagement.

> *Note:* The procedures in this chapter are written for right-handed Marines; left-handed Marines should reverse directions as needed.

Shooting on the Move

When moving from one area of cover to another, the Marine may engage a target that presents an immediate threat. In this situation, the Marine relies on accurate fire to serve as his cover.

Moving With the Pistol

The likelihood of encountering a threat dictates how the pistol is carried while the Marine is moving. But regardless of the method of carry, the pistol's muzzle is pointed in the same direction that the head and eyes are looking—the eyes and muzzle move as one.

When carrying the pistol in the Alert position, the pistol is tucked in close to the body, finger off the trigger. This position allows the Marine freedom of movement and still allows for a quick presentation of the pistol.

If there is a high probability of encountering a threat, the Marine carries the pistol in the Ready position while moving.

Moving: The Glide Technique

To shoot accurately while moving, the Marine cannot run or walk fast. A normal running movement produces too much bounce in the Marine's body and makes it extremely difficult to use the sights of the pistol or to achieve the needed level of stability for accurate shooting. Therefore, the proper movement technique is similar to a glide and allows for accurate target engagement:

- A lower center of gravity is achieved by keeping the knees bent and the upper body erect.
- The bent knees also serve to absorb the shock generated by movement.
- The feet are not lifted as high as in a normal walk or run, allowing obstructions to be kicked out of the way.
- The feet and knees point in the direction of travel. If there is a need to engage a target off to the side, the Marine rotates at the waist to move the upper body in the desired direction. The upper body shooting position does not change.

The glide is not a technique that can be used for a long period of time or for long distances. Rather, the Marine can move in a normal manner until it becomes necessary to engage a target, he then slows and assumes the glide.

Engaging Targets: Using the Pistol Sights

When the Marine fires the pistol, he must concentrate on the pistol's front sight. The adherence to the fundamentals of marksmanship becomes even more vital when both the Marine and the enemy may be moving.

Continuing to Move

A Marine continues to move no matter what happens. Continuing to move makes it difficult for the threat to engage the Marine and can be one of the Marine's best defensive assets. When moving to cover or to a different position, the Marine should not stop to engage a target.

Reloading and Stoppages

If a stoppage occurs or a reload is required while moving, the Marine's primary consideration is to keep moving and seek cover. Ideally, reloads and clearing stoppages are performed behind cover.

Turn and Fire

The techniques for turn and fire allow a Marine to engage a target that is not directly in front of him. The key to turn and fire is smoothness and quickness of pivoting and presentation of the pistol to engage the threat. Turn and fire is based on the principle that wherever the head goes, the body follows.

Engaging Targets 90 Degrees to the Right and Left

The Marine turns the head toward the threat and identifies the target. Once the target is identified, the Marine maintains focus on the target for the rest of the presentation.

If engaging a target 90 degrees to the right, the Marine turns toward the target by raising the left foot while pivoting on the ball of the right foot. The Marine plants the left foot once he is facing the target squarely. The stance is natural. See figure 9-1. The Marine then engages the target.

Figure 9-1. Engaging Targets 90 Degrees to the Right.

If engaging a target 90 degrees to the left, the Marine turns toward the target by raising the right foot while pivoting on the ball of the left foot. The Marine plants the right foot once he is facing the target squarely. The stance is natural. See figure 9-2. The Marine then engages the target.

Engaging Targets 180 Degrees to the Rear

The Marine turns the head toward the target, looking over either the right or left shoulder, and identifies the target. Once the target is identified, the Marine maintains focus on the target for the rest of the presentation. See figures 9-3 on page 9-4.

If the Marine pivots to the right to engage the target, pick up the left foot, move it across the right foot, while pivoting 180 degrees on the ball of the right foot, plant the left foot, and squarely face the target in a natural stance. See figure 9-4 on page 9-4.

If the Marine pivots to the left to engage the target, pick up the right foot, move it across the left foot, while pivoting 180 degrees on the ball of the left foot, plant the left foot, and squarely face the target in a natural shooting stance. See figure 9-5 on page 9-4.

The Marine then engages the target.

Figure 9-2. Engaging Targets 90 Degrees to the Left.

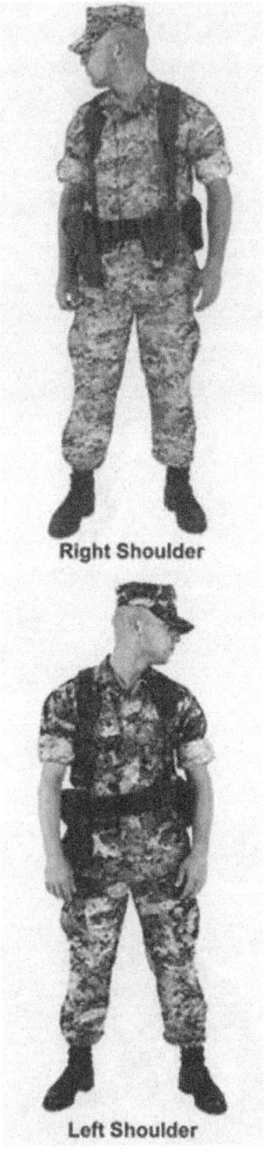

Figure 9-3. Engaging Targets 180 Degrees to the Rear.

Figure 9-4. Engaging Targets 180 Degrees to the Rear (Right Shoulder).

Figure 9-5. Engaging Targets 180 Degrees to the Rear (Left Shoulder).

CHAPTER 10
LOW LIGHT AND DARKNESS TECHNIQUES

An effective combat marksman must be prepared to detect and engage targets under a variety of conditions. Factors such as terrain and opportunity often dictate that the Marine engage combat targets at night or under low-light conditions.

The fundamentals of marksmanship are employed for engagement of targets in low light and darkness just as they are in daylight. However, the principles of night vision and target detection when engaging targets at night must be applied.

Combat Mindset

In the stress of the combat environment, Marines must eliminate any hesitation, fear, or uncertainty of action and focus on the actions to fire well-aimed shots. This is important during low light and darkness when attention is more easily diverted because the sense of vision is reduced. This may create a shock of awareness as the Marine relies more heavily on his other senses. Every noise, movement, and muzzle blast appears intensified at night. The physical acts of shooting must become second nature to the Marine, so his focus will not be diverted from firing well-aimed shots. This is accomplished through mental preparedness and training until shooting actions become instinctive.

Night Vision Adaptation and Maintenance

Night Vision Adaptation

There are two methods for acquiring night vision.

The first method is to remain in an area of darkness for about 30 minutes. This area can be indoors or outdoors. The major disadvantage of this approach is that a Marine is not able to perform any tasks while acquiring night vision in total darkness.

The second method is to remain in a darkened area under a low intensity red light (similar to the light in a photographer's darkroom) for about 20 minutes, followed by about 10 minutes in darkness without the red light. This method produces almost complete night vision adaptation while permitting the performance of some tasks during the adjustment period.

Night Vision Maintenance

Because the eyes take a long time to adjust to darkness, it is important to protect night vision once it is acquired. To maintain night vision, the Marine avoids looking at any bright light and shields the eyes from parachute flares, spotlights, or headlights. When using a flashlight to read a map or other material—

- Put one hand over the glass to limit the area illuminated and the intensity of the light.
- Keep one eye shut to reduce the amount of night vision lost.
- Cover the light with a red filter to help reduce the loss of night vision.

Target Detection Techniques

Once the Marine has acquired night vision, he is prepared to locate targets. Some of the daylight observation techniques, such as searching for target indicators, also apply at night or in low light. But night observation techniques must allow for the limitations of night vision and the need to protect it.

Off-Center Vision

Because of the placement of the cones in the center of the retina and the rods around the edges, the angle at which the Marine observes an object at night affects how well it is seen. Off-center vision is the technique of keeping the attention focused on an object without looking directly at it. To maximize the use of the eye's rods that provide night vision—

- Never look directly at the object being observed.
- Look slightly to the left, right, above, or below the object.

Each Marine must experiment and practice this technique to find the best off-center angle. Typically, the best off-center angle is 6 to 10 degrees away from the object or about a fist's width at arm's length. See figure 10-1.

Scanning/Figure Eight Scan

Scanning is the use of off-center vision to observe an area or object and involves slowly moving the eyes in a series of separate movements across the objective area. See figure 10-2. To execute a figure eight scan, move the eyes in a figure eight pattern in short, abrupt, irregular movements over and around the area. Once a target indicator has been detected, focus should be concentrated in that area, but not directly at it. It is more effective to scan from a prone position or a position closer to the ground than the object being observed, creating a silhouetted view of the object. When scanning an area, look and listen for the same target indicators as in daylight: movement, sound (which seems louder at night than during daylight), and improper camouflage. Also, objects in bright moonlight/starlight cast shadows just as in sunlight.

Effects of Illumination

Both ambient (natural) light and artificial illumination can affect a Marine's perception of the target's distance and size and his night vision. The following situations produce less of an effect at the close ranges that pistol engagements occur, but they still affect target detection and engagement:

- Light behind or between the Marine and the target illuminates the front of the target and makes it appear closer than it actually is.
- Light beyond the target displays the target in silhouette and makes it appear farther away than it actually is.
- The introduction of light requires the eyes to make a sudden, drastic adjustment to the amount of light received, causing a temporary blinding effect that occurs when night vision is interrupted abruptly. Ambient light can also have the

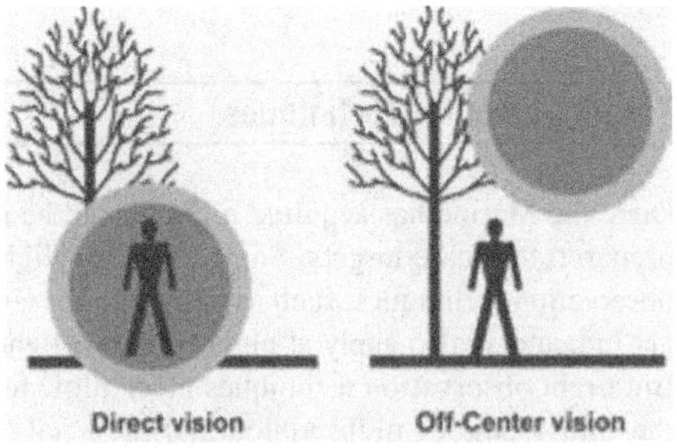

Figure 10-1. Off-Center Vision.

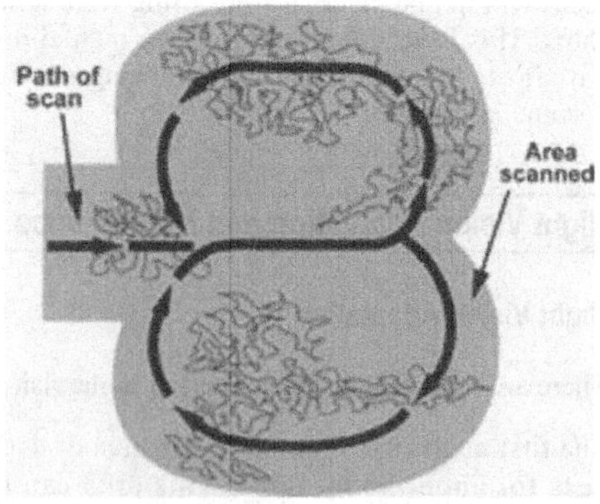

Figure 10-2. Figure Eight Scan.

same blinding effect; e.g., when a bright moon suddenly appears from behind clouds.

Acquiring Targets at Night

The Marine must keep both eyes open to get maximum visual coverage of the target area. Keeping both eyes open also improves depth perception and peripheral vision.

Sight Alignment/Sight Picture

Sight alignment and sight picture are obtained the same way in darkness as in daylight. There is normally enough ambient light to perceive objects as far away as 50 meters, especially if they are moving. However, when pistol sights are placed on a dark background, such as a camouflaged target, the Marine may not be able to acquire and align the sights clearly. Instead, the Marine may have to rely almost entirely on his presentation, firing position, and grip to get the pistol on target. To check for sight alignment or acquire the sights:

- Move the pistol just off the edge of the target to an area that provides a good contrast.
- Acquire sight alignment.
- Bring the sights back on line with the target while applying pressure to the trigger so the shot breaks once the sights are on target.

Flashlights

A flashlight can be used in low light and darkness to identify and illuminate targets. It is also used to acquire sight picture.

Types of Flashlights

There are a variety of flashlights used throughout the Marine Corps. They come in many different shapes and sizes; however, they are either straight/tubular or L-shaped.

Straight/Tubular

The straight/tubular flashlight is the most common. The body of the flashlight is a straight/tubular shape with the lens at the head of the flashlight. The on/off mechanism is located along the body or at the base of the flashlight.

L-Shaped

The L-shaped flashlight has a straight/tubular body, with the head/lens of the flashlight positioned at a 90-degree angle to the body of the flashlight. The on/off mechanism is located along the body or at the base of the flashlight.

Activation Devices

Depending on the type of flashlight, there are several ways that it can be activated.

On/Off Button

On/off buttons are located either along the body of the flashlight or at the base. On/off buttons are operated in a number of ways, depending on the flashlight.

On/Off Switch

On/off switches are mounted on the body of the flashlight. This switch has three settings that allow the flashlight to be either on, off, or in manual mode. In the manual mode, the flashlight is turned on by pressing and holding the button located next to the on/off switch.

Rotating Head

Rotating the head of the flashlight, either clockwise or counterclockwise, turns the flashlight on. Continuing this motion allows the width and intensity of the beam to be adjusted.

Types of Lenses

Clear Lenses

Clear lenses are used primarily for target detection and to illuminate the pistol sights. They are the most commonly used lens.

Colored Lenses

Colored lenses include red, amber, and blue. These lenses are used primarily for map reading and signaling, but can be used to illuminate the pistol's sights.

Target Detection

When a target is illuminated, the front sight may become silhouetted against the target, providing the sight picture needed to engage the target. During target detection, the Marine's focus should be twofold: to scan the area to identify possible targets and to assess the area to formulate a plan for engagement or cover (e.g., identifying the quickest route to cover, determining the best method for engagement based on terrain).

Searches

The Marine executes a hasty search immediately upon entering a new area in order to identify a threat that poses immediate danger. To execute a hasty search—

- Quickly scan the area with the flashlight taking note of obvious points throughout the area that could cover or conceal the enemy. Wherever the eyes move, the flashlight should move.
- Aim the flashlight beam on the ground about 8 to 10 feet in front of the Marine's location, this allows the eyes to follow the beam and quickly establishes a reference point for the light. Aiming the beam at the outer edge of the search area strains the eyes to find the beam and decreases the field of view.

The Marine executes a detailed search after the hasty search. A detailed search focuses on target indicators identified during the hasty search. To conduct the detailed search—

- Focus the most direct or intense portion of the flashlight beam on the target indicators.
- Move the beam of light slowly across the target indicator from right to left or left to right. Wherever the eyes move, the flashlight should move.

Considerations

During night operations, the Marine should use a flashlight whenever possible. A flashlight not only helps the Marine locate targets at night, but the beam can be aimed directly at the target once it is detected and temporarily blinds the target. This gives the Marine the advantage to react before the target does. However, the Marine must consider the following:

- Anytime a flashlight is on, the location may be revealed to the enemy. The Marine should keep the flashlight pointed out in front of his body to help avoid illuminating himself and revealing his position.
- Light shined directly from in front of the Marine at the target obscures the Marine.
- Light shined from the side of the Marine at the target illuminates the Marine.
- Light can bounce off surfaces and reflect back onto larger areas that the light is not focused on. Therefore, whenever a Marine shines a flashlight onto a surface, some of the light may reflect back and illuminate the Marine.
- When positioned at the outside corner of a room, building, or other cover, the Marine should avoid pointing the flashlight beam directly at the corner because the beam reflects off the corner and illuminates the Marine. To prevent this, the head of the flashlight should extend just beyond the corner.
- Many flashlights have adjustable beams that aid in target detection. The beam's intensity must be adjusted to provide the best illumination of the area depending on the distance between the Marine and the area of observation.
- Diffused light from a wide beam creates a softer light and illuminates a greater area, but the beam does not travel a great distance. A wide beam is best for observing larger areas at close range.
- Concentrated light from a narrow beam illuminates a smaller field of view, but the beam travels a greater distance. A concentrated beam is effective for observing a specific area or an area that is further away. Concentrated light

blinds the enemy and prevents him from focusing on the Marine or determining his location.

- When searching an area, the Marine can alter the position where he is holding his flashlight so that the beam is perceived as coming from various locations. For example, the Marine can turn the flashlight on from a standing position, quickly search for targets, turn the flashlight off, assume a kneeling position, and search again. This keeps the enemy from obtaining an exact location of the Marine's position.
- Keep both eyes open to get maximum visual coverage of the target area. Keeping both eyes open also improves depth perception and peripheral vision.

Target Engagement

Once a target is detected using a flashlight, the flashlight can be used to acquire sight picture and facilitate engagement of the target. The fundamentals of marksmanship are employed for engagement of targets in darkness just as they are in daylight. However, the Marine must apply the principles of night vision and target detection and must be able to employ a flashlight properly when engaging targets at night.

Grip

To engage a target accurately while using a flashlight, the flashlight must be held to maintain the pistol's stability, control, and recoil during firing. Normally, the left hand (two-handed grip) provides the pistol's stability and ability to manage recoil. However, when firing with a flashlight, the left hand holds and operates the flashlight. Therefore, some stability, control, and management of recoil is lost.

Securing the Flashlight

Most flashlights come equipped with a retaining loop located at the base for ease of transportation. To position the cord properly:

- Feed one end of the cord through the retaining loop of the flashlight and tie the ends of the cord together.
- Slip the left hand through the cord loop. The loop should be just big enough so that, if the flashlight needs to be dropped, it can be retained on the wrist.
- Rotate the flashlight until the cord is twisted to the desired tension and length necessary to best stabilize the rear end of the flashlight in the hand.
- Grasp the flashlight with the left hand in a position that allows it to be operated easily.

If the flashlight has a retaining loop, attaching a cord to the loop helps support and stabilize the flashlight when the Marine must fire the pistol with the flashlight in his left hand. The cord also helps retain the flashlight.

Acquiring Sight Alignment/Sight Picture

When holding a flashlight on a target, the tendency is to look at the target rather than the sights. Sight alignment is still necessary for effective target engagement. Using a flashlight to illuminate a target allows the Marine to—

- Acquire sight picture by silhouetting the sights against the target.
  - The Marine establishes sight picture by focusing the sights in the soft, diffused light area of the target.
  - At close ranges, colored lenses produce enough light on the target to silhouette the sights; at long ranges, colored lenses will not provide enough light off the target to illuminate the pistol sights. The Marine must be able to see the target by some other means, (e.g., contrast, moonlight).
- Acquire sight picture by illuminating the sights.
  - A colored lens can be held directly over the sights to shine light on the top of the pistol to illuminate the sights. This allows sight alignment to be established without revealing the Marine to the enemy. This method can also be used with a clear lens to make a precision shot at long ranges but the Marine risks being illuminated by the light.
  - A colored lens can be shined from directly behind the pistol sights to illuminate them. This

allows sight alignment to be established, but reveals the Marine's position to the enemy.

Target Engagement Techniques

Two-Handed Grip Technique. This technique is typically used with a straight/tubular flashlight. The pistol and flashlight must be side by side and level so the Marine can engage a target without making adjustments to the pistol or the flashlight. This technique is also effective for firing multiple shots because the flashlight and pistol recoil as a unit.

Whenever possible, the flashlight is just in front of the muzzle of the pistol so that it does not illuminate the pistol. Placement of the flashlight alongside the pistol may need to be adjusted depending on the size of the flashlight or the location of the on/off mechanism.

This technique is most often used with an Isosceles position (see fig. 10-3):

- Grasp the flashlight with the left hand, lens pointing down range.
- Wrap the thumb and index finger of the left hand around the body of the flashlight, thumb resting on the on/off switch to allow easy operation of the on/off switch without disrupting the grip on the flashlight.
- Extend both arms toward the target and bring the flashlight alongside the pistol so the fingers of the left and right hands touch.
- Wrap the bottom three fingers of the left hand around the fingers of the right hand, incorporating the flashlight into a two-handed grip.
- Apply isometric pressure against both hands to aid in stabilizing the pistol.

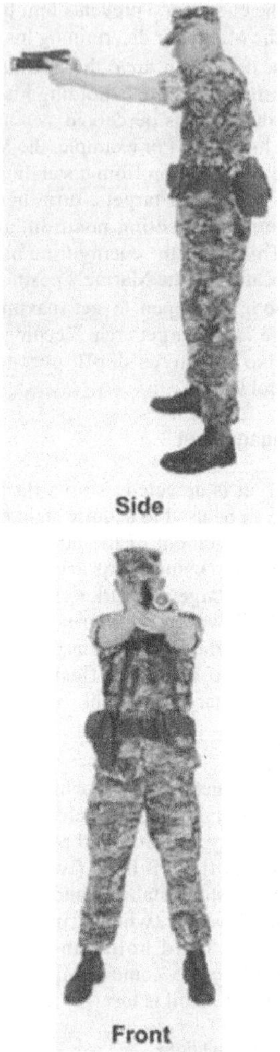

Figure 10-3. Two-Handed Grip Technique With Flashlight (Side and Front View).

This technique can also be incorporated into a Weaver position; the more the left hand wraps around the right, the more the body can be angled and the left elbow bent. Apply push-pull pressure on the grip to stabilize the pistol. See figure 10-4.

Cross-Hand Technique. The cross-hand technique is used primarily with a Weaver position and can be used with either a straight/tubular flashlight or an L-shaped flashlight (see fig. 10-5):

- Grasp the flashlight with the left hand, fingers wrapped around the top of the flashlight and thumb wrapped around the bottom.

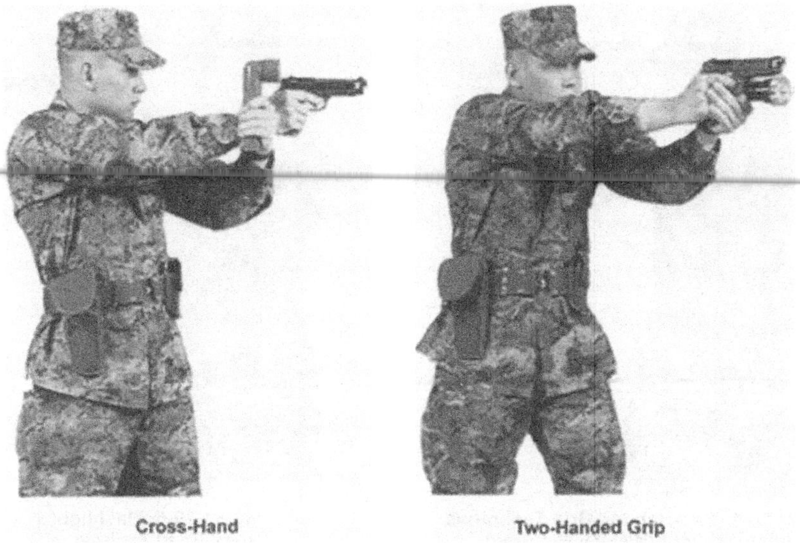

Figure 10-6. Flashlight Ready Carry (Cross-Hand and Two-Hand Grip).

Figure 10-7. Flashlight Alert Carry (Two-Handed Grip and Cross-Hand).

Figure 10-4. Two-Handed Grip Technique With Flashlight (Weaver Position).

Figure 10-5. Flashlight Cross-Hand Technique.

- Place the thumb on the on/off button while maintaining a firm grip on the flashlight.
- Extend both arms toward the target and bring the left hand under the pistol so that the back of the left hand is resting firmly against the back of the right hand.
- Apply isometric pressure against both hands to aid in stabilizing the pistol.
- Extended the right arm fully and bend the left arm at the elbow.

Considerations for the Carry/Transport

When searching an area for targets, the Marine moves with the pistol at the carry or transport dictated by the threat level.

The Ready. If enemy contact is expected (contact imminent), the Marine carries the pistol at the Ready and carries the flashlight in his left hand and incorporates it into his firing grip. In the Ready, the Marine can search for targets and readily present the pistol and flashlight to the target for engagement without making adjustments to either the grip or the flashlight. See figure 10-6 on page 10-8.

The Alert. If enemy contact is likely, the Marine carries the pistol at the Alert and the flashlight in his left hand and incorporates it into his firing grip. The Marine must ensure the flashlight is not pointed at the deck because it illuminates the Marine. If necessary, the head of the flashlight may have to be tilted up to elevate the beam to increase the field of view. If the flashlight has been tilted, it has to be lowered so it is level with the pistol muzzle and the light shines directly on the target in order to establish sight picture. See figure 10-7 on page 10-8.

Holster Transport. If there is no immediate threat, the pistol is holstered and the Marine carries the flashlight with his left hand to search for targets. Should a target present itself, the Marine presents the pistol from the holster and engages the target. Time and distance to the target dictate whether the Marine incorporates the flashlight and his left hand into the firing grip.